THE ECONOMICS OF BUSINESS CULTURE

The Economics of Business Culture

Game Theory, Transaction Costs, and Economic Performance

MARK CASSON

CLARENDON PRESS · OXFORD

Oxford University Press, Walton Street, Oxford OX2 6DP

Oxford New York Toronto
Delhi Bombay Calcutta Madras Karachi
Kuala Lumpur Singapore Hong Kong Tokyo
Nairobi Dar es Salaam Cape Town
Melbourne Auckland Madrid

and associated companies in
Berlin Ibadan

Oxford is a trade mark of Oxford University Press

Published in the United States
by Oxford University Press Inc., New York

British Library Cataloguing in Publication Data
Data available

Library of Congress Cataloging in Publication Data
Casson, Mark, 1945–
The economics of business culture : game theory, transaction costs, and economic performance/Mark Casson.
p. cm.
Includes bibliographical references and index.
1. Corporate culture. 2. Business ethics. 3. Game theory.
4. Decision-making, Group. I. Title.
HD58.7.C37 1991 338.5'14dc20 91–2067
ISBN 0–19–828375–X

3 5 7 9 10 8 6 4 2

Printed in Great Britain
on acid-free paper by
Bookcraft (Bath) Ltd
Midsomer Norton, Avon

For Terry Henderson

Preface and Acknowledgements

The idea for this book has been gestating for a long time, beginning with my dissatisfaction with my earlier work on *The Entrepreneur*. This work failed to consider the leadership qualities of the entrepreneur, and in particular the way that the personality of the founder can exert a cultural influence within the firm. On closer examination, it became evident that entrepreneurs themselves are strongly influenced by their own cultural environment. They intermediate not only in the production and exchange of goods but also in the transmission of cultural values. These values are developed principally within religious, ethnic, and national groupings, and seem to exert a major influence on the economic performance of these groups. The economic analysis of culture should therefore be able to shed light on a wide variety of contemporary social and business problems. The object of this book is to develop the relevant analysis and to substantiate this claim.

The book is aimed at a fairly wide audience and is written so that, as far as possible, it can be read both at the intuitive and at the technical level. The technical demands are fairly modest—principally an acquaintance with the calculus of a single variable. Some of the most commonly occurring technical terms are explained in the glossary at the end of the book. Readers who feel that they are becoming 'bogged down' in technical detail can always turn to the summary at the end of the chapter, which reviews the major results in a non-technical way. It is possible, in principle, to follow the gist of the deductive arguments simply by reading Chapters 1, 13, and 14, and the introductions and summaries to the chapters in between.

A first draft of this book was written over the Christmas vacation 1988 and a second draft over the Christmas vacation 1989. I am grateful to my wife Janet and daughter Catherine for their toleration of this anti-social behaviour (which, incidentally, violates the moral principles referred to in this book). Janet also read through the final draft and made suggestions for improvement, with the interests of the non-specialist reader in mind.

My early ideas were refined in the course of conversations with Peter Buckley, John Creedy, Geoff Jones, Sir Arthur Knight, Robert Locke, Steve Nicholas, and John Shannon, and with my former parish priest Terry Henderson, to whom this book is dedicated. Emerging hypotheses were initially applied in the international business field in *Enterprise and Competitiveness* and, when generalized, were expounded in seminars at Bristol, Reading, Melbourne, La Trobe, London Business School, and the Industrial Economics Study Group meeting in London. I am grateful to the participants in these seminars for their toleration of a somewhat eccentric paper. The publisher's anonymous readers also provided extremely valuable comments. Finally, I wish to thank Jill Turner for once again typing up my work with such speed and efficiency.

M.C.C.

Contents

List of Tables x

I. Principles of Leadership 1

1. Why Culture Really Matters 3
2. How Leaders Inspire Achievement 29

II. Co-ordination with Pairwise Encounters 53

3. How Culture Sustains Trade 55
4. The Causes of Catastrophe 82
5. Promoting Participation 100
6. Reciprocity and Revenge 116

III. Co-ordinating Work-Groups 131

7. Team Spirit 133
8. Intermediators: The Middle-Class Middlemen 148
9. Small is Cosy: Intimate Relations in Small Groups 169

IV. Collective Co-ordination and Social Justice 187

10. From Free-riding to Philanthropy 189
11. Distribution and Justice 201
12. Collusion as Moral Crusade 213

V. Synthesis 223

13. Business Culture 225
14. Policy Implications 243
15. Summary and Conclusions 255

Notation 264
Glossary 266
Bibliography 269
Index 275

Tables

2.1 Data for follower's effort choice 32
2.2 Objectives of an altruistic leader 34
2.3 Comparative statics: sensitivity of the optimal values 40
2.4 Comparative statics: qualitative results 40
2.5 Impact of leader-, group-, and institution-specific factors on the optimal intensity of manipulation 41
2.6 A comparison of the optimization strategies associated with the four leadership objectives identified in Table 2.1 43
2.7 Distortions created by substituting narrow materialism for alternative leadership objectives 45
2.8 Follower's data set under monitoring 47
3.1 Material rewards from an encounter 59
3.2 Emotional rewards from an encounter 60
3.3 Follower's data set for an encounter 60
3.4 Follower's data set for a trading encounter 63
3.5 Impact of the intensity of manipulation on the crime rate under a uniform bivariate distribution of sensitivity and confidence 69
3.6 Interior optima for the three regimes exhibiting diminishing returns to manipulation 72
3.7 Comparative statics of the interior optima: the intensity of manipulation and the crime rate 73
3.8 Comparative statics of the interior optima: the gains from manipulation 75
3.9 Transitions between regimes induced by an increasing intensity of manipulation 76
3.10 Announcement strategies and their value 78
4.1 Dynamics of the crime rate with repeated operation 94
5.1 Trader's data set 102
5.2 Follower's data set with voluntary participation: the general case 108
5.3 Strategic outcome in $g - p$ space 110

5.4 Equilibrium and stability of recurrent trading with voluntary participation 114
6.1 Modified emotional rewards 118
6.2 Modified follower's data set 118
6.3 Follower's information set with feelings of satisfaction from reciprocity 122
6.4 An encounter as a three-stage game 126
6.5 Rewards associated with revenge 126
6.6 Modification of overall rewards when an honest victim is known to take revenge on a cheat 127
7.1 Member's data set for team production 138
8.1 The process of intermediation 152
8.2 Information sets for customer and middleman under trade 152
8.3 Information sets for customer and middleman: manipulation by middlemen who may cheat 155
8.4 Information sets for follower and middleman: monitoring by middlemen who may cheat 157
8.5 Information sets for team member and team leader 163
9.1 Follower's perceived rewards for participating in trade under the guilt mechanism 171
9.2 Average costs of alternative communication strategies, by size of group 177
9.3 Average cost of alternative communication strategies for monitoring, by dispersion of group 178
9.4 Average cost of alternative communication strategies for manipulation, by dispersion of group 178
9.5 Least-cost communication strategies 178
10.1 Follower's data set for public-good free-rider decision problem 192
10.2 Follower's data set for private-good free-rider decision problem 195
11.1 Data for follower's choice when subject to lump-sum tax 205
12.1 Information set for member of a cartel 215
12.2 Distributional effects of cheating in a cartel 215
13.1 Functions of groups 226
13.2 A typology of groups 227

PART I

Principles of Leadership

1

Why Culture Really Matters

1.1. Introduction

This book has a simple point to make. Overall economic performance depends on transaction costs, and these mainly reflect the level of trust in the economy. The level of trust depends in turn on culture. An effective culture has a strong moral content. Morality can overcome problems that formal procedures—based on monitoring compliance with contracts—cannot. A strong culture therefore reduces transaction costs and enhances performance—the success of an economy depends on the quality of its culture.

The point is not new. It has been made many times by sociologists, anthropologists, social psychologists, and historians, not to mention politicians and preachers. The point seems to be new to many economists, though, who have blinded themselves to the obvious by adopting a narrow view of human nature. With certain minor modifications to conventional economic modelling techniques, however, a broader view incorporating culture can easily be developed, and this book shows how it can be done.

Were this book being written 150 years ago its message would have been an optimistic one—explaining how moral regeneration could improve economic performance. Today a more pessimistic note may be in order—the book also spells out the catastrophic consequences of moral decline. It highlights the fact that there is a critical level of mutual trust in an economy, which is necessary to prevent it from disintegrating into anarchy.

1.2. Comparative Cultural Analysis

The influence of trust on economic behaviour is ubiquitous (Leff

1986). One of the simplest ways of appreciating this is by paired comparisons of two extreme situations. Three such comparisons are offered below. They are concerned with personnel policy, inter-firm co-operation, and urban growth and decline. In each case the effects of a high-trust culture and a low-trust culture are compared. Although the comparisons show the high-trust culture in a favourable light, the aim is not to demonstrate its universal superiority. The function of these comparisons is rather to highlight the very large opportunity costs that can be incurred when an economy becomes trapped in a low-trust culture. Although a low-trust culture is sometimes efficient in material terms, it is a common mistake for those who exploit it to believe that it provides the only basis on which economic activity in general can be organized.

Personnel policy

Consider two production plants operated with contrasting management styles. Plant A has a manager who does not trust his workers. He feels obliged to supervise them continually in case they slack. To make supervision easy the manager has de-skilled the work as much as possible (in line with 'Fordist' and 'Taylorist' thinking). He can tell at a glance whether someone is working hard or not.

The manager knows that the workers do not enjoy their work, and he pays a premium on the normal wage to compensate for this. Workers join because they need to earn the money, often for some specific use, such as buying a house. When they have earned sufficient, they usually quit because they cannot stand the stress induced by the fast pace and the monotonous work. No one wants to make a career in the plant, and so turnover is high. The manager is aware of this too.

He also knows that long-term contracts attaching workers to the firm are unenforceable. He is therefore quite unwilling to finance training. He may occasionally finance a small amount of job-specific training, though he would rather de-skill the work further to avoid this if possible. But he will certainly not finance general training because this would only help the worker to get another job. The worker cannot finance the training himself: he cannot pledge his future labour to the bank as collateral for a loan, and the main reason why he is working for the firm in the first place is that his personal wealth is limited. Workers therefore receive practically no training at all.

Workers without general training know that it will be difficult to get another job that pays as well. The manager knows this and exploits it to the full. He continually reminds the workers of how easily he can fire them if they are caught slacking. He may even choose to locate his plant in an area of high unemployment and select employees who have heavy family commitments in order to maximize the force of this threat.

Because the workers are untrained they are also inflexible. If there is a recession wages are cut and when the statutory (or nationally negotiated) minimum is reached they are simply fired. The manager adjusts to long-term structural change by a similar mechanism—firing one lot of workers and hiring another lot with a different set of aptitudes (e.g. firing older workers and hiring school-leavers if more dextrous work is involved). If a shortage of workers develops then the manager will lobby for the government to relax immigration requirements, reduce taxes on labour income to promote participation, and lower the school-leaving age.

The manager is, of course, strongly anti-union. His low-training strategy means that he relies on low price rather than quality of workmanship to sell his output. A union monopoly threatens price competitiveness not only in respect of the wage rate, but most importantly in respect of the speed of the production line. Attempts to slow the line directly threaten his 'sweated labour' policy. He has enough problems with 'spanner in the works' sabotage of the line as it is.

But above all, where unions are concerned, he fears something that he does not understand—the power of union leaders to build solidarity among their members. This solidarity undermines the manager's 'divide and rule' strategy of firing slackers. The other workers strike in sympathy if one of them is sacked unjustly. The manager cannot understand why workers that are in need of money will sacrifice it through an apparently pointless gesture of solidarity. The idea that participation in such a gesture can confer a major emotional benefit is totally alien to him.

In plant B, by contrast, the manager trusts his workers. He makes it known that he expects them to work hard when they are not being supervised. Supervisors act more as counsellors and consultants to the workers than as policemen. Workers like the plant, even though the pay is not particularly good, because it has a friendly atmosphere. If they have an urgent need for cash, the firm may be able to help by offering overtime for a limited period. Labour turnover is low and the firm can therefore afford to offer training of both a general and a specific kind. The manager knows that workers are loyal and will not threaten to quit because their potential earnings have been raised by training undertaken at the firm's expense.

The firm reinforces workers' loyalty with an implicit contract that keeps wage rates and employment stable during a recession—though possibly a work-sharing scheme may be used to shorten the working week if the recession is severe. The manager adopts a 'holistic' approach towards industrial relations, which means that he is heavily involved in local community affairs as well. The production teams within the plants are microcosms of the local community from which the workers are hired.

The manager's familiarity with the principle of community means that he is happy for them to join a union. The manager trusts them not to be trouble-makers. The workers show little interest in the union, however, except as a wage-negotiating agent and as a friendly society providing insurance through its welfare fund. In wage negotiations both the manager and the union accord considerations of equity as well as cost-efficiency a prominent role.

Because the workers have a general as well as a specific training, they are not only confident in their job but also flexible. Jobs do not, therefore, need to be de-skilled. Neither do roles need to be tightly specified, because workers, being confident, are happy to give advice, and to take it from others. The firm responds to structural change by keeping the workers together as a team, but redeploying them to a different kind of work. Workers have sufficient mental flexibility to remain with the firm even if it diversifies into a quite different industry.

In international competition plants of type A are best adapted to operating large-scale manufacturing processes where machinery is used to de-skill the work. The close proximity of workers within the plant affords economies of supervision—a single supervisor can readily oversee a large number of operatives. A long production run is advantageous too, because it requires little flexibility from the work-force once start-up has been achieved. (With inflexible workers the learning curve starts high and declines only gradually, so that the economies of de-skilling are achieved only after a fairly long run).

Large-scale, long production runs, and price- rather than quality-competitiveness are all characteristic of mature manufactured products. The type B plant, by contrast, is best suited to the small-scale production of innovative differentiated products. Continual innovation creates a succession of short production runs which maintain the interest and motivation of flexible workers.

Type B plants are appropriate to the service sector. This is particularly true of professional services, where it is very difficult to monitor the quality of work through supervision. The subordinate is often more expert than his superior, and so quality is maintained by encouraging the subordinate to respond positively to the trust that is placed in him.

In practice, firms may combine these two management styles. A core of permanent staff is used to manage innovation and start-up and to maintain expensive equipment. Routine process work is carried out by non-tenured staff, who are hired and fired as demand conditions dictate. The permanent staff constitute an élite who socialize regularly amongst themselves, whilst the 'rank and file' production workers are kept on the social periphery.

Inter-firm co-operation

In country A the managers of different firms trust each other but in B they do not. The managers in A believe that it does not matter that they do not trust each other, however. Country A has a very explicit constitution which safeguards the rights of the individual, and makes it easy for victims of commercial malpractice to sue each other. Everyone in A checks with his lawyer before signing any agreement, and then relies on the law to enforce it. Contracts are complex, but the damages awarded for breach of contract are correspondingly high.

In country B the managers trust one another. They are quite bewildered by the legal formalities they encounter when they try to do business with A. The managers in B come to distrust those in A because they assume that there must be some catch in the contract if the other party is so anxious they should sign. Conversely the managers in A do not trust those in B—partly because they do not trust anyone anyway, but more particularly because they interpret reluctance to sign as reflecting lack of commitment to deliver the goods. Trade is impaired, and gains from comparative advantage and large-scale production are lost as a result.

When a manager in A needs to guarantee the quality and delivery dates of raw materials or component inputs he finds that even complex contracts cannot cover all the things that could possibly go wrong. He really needs control over supplies himself, and so he integrates backwards, say by taking over his supplier's business. The manager in B tackles the same issue in a different way—by developing a close personal relationship with his supplier. In B the supply contract is deliberately vague, so that both parties can have the flexibility to improvise solutions to problems as and when they occur. It is deemed unnecessary to plan in detail for highly improbable contingencies, because each can trust the other to help out. Solutions are developed collaboratively, not just at the office, but on the golf course or over lunch as well.

When two managers in A find that their firms possess complementary technologies their lawyers have to be called in to negotiate complex cross-licensing agreements. But the negotiations typically founder on legal technicalities, and in frustration one of the firms simply buys out the other. Some of the other firm's key scientists then leave, because of some snub to their professional status which the acquiring firm's managers simply cannot understand. As a result, the full potentiality of the complementarities is never realized.

Two managers in the same position in B quickly form a joint venture. The ambiguity of control involved in a 50:50 joint venture does not bother them since they trust each other. Indeed, they welcome the fact that each is hostage to the other, since by deepening their trust the new relationship will stimulate further collaboration.

In country A collaboration tends to be regarded as a sign of weakness. Surely a technological leader must be self-sufficient, and wish to keep his secrets to himself? Only the second- and third-best will wish to co-operate—co-operation merely signals their inferiority. In country B, by contrast, the ability to collaborate is seen as a source of strength. Collaboration extends your sphere of influence, and a reputation as a good collaborator means that people will bring their ideas to you in order to involve you in their exploitation.

Managers in A mainly rely on State education to train the skilled workers and professionals needed in their business. They do not collaborate to fund vocational training. Because they provide no funding, they have only limited influence over the curriculum. Managers in B, however, find it relatively easy to collaborate over training. A 'no poaching' code, upheld by custom, means that firms cannot consistently recruit more trained people from other firms than they have lost to them. Thus everyone must recruit new trainees in proportion (roughly) to the size of their work-force. There is therefore a strong incentive to contribute to a training 'club'. In recognition of the club subscriptions paid by employers, the government offers counterpart funds, even though the curriculum is very much under employers' control. In A, by contrast, the government resents the political unpopularity of the taxation required to fund vocational training, and so provides a training of rather limited quality so far as employers are concerned.

Overall, economy A relies on a small number of relatively rigid organizational structures—principally firms and government—linked by impersonal 'arms-length' markets. Economy B, on the other hand, comprises a rich diversity of arrangements—ranging from open-ended long-term contracts through joint ventures to inter-firm clubs. Even the dividing line between firms and government is sometimes hard to discern in B. To the outside observer it may be difficult to determine exactly how co-ordination in B is achieved. The difficulty arises partly from a misguided attempt to search out a new kind of organizational structure to be found only in B. There is no single structure of this kind. The competitive strength of B lies in the sheer diversity of organizational forms that can be used. This in turn reflects the flexibility afforded by a high-trust culture in the innovation of new contractual arrangements.

Urban growth and decline

City A developed many years ago as an entrepôt port and has since become a major financial centre. In the late nineteenth century many migrants stopped off here instead of proceeding onward with their journey, and the legacy of continued migration since then is a varied collection of ethnic communities. Each community occupies a particular part of the city and is relatively closed to outsiders. The professional élite live out in the suburbs, fringing on the countryside or the coast, and commute along railways and motorways that pass between the inner city tenements where the poorer communities reside.

Tourists come to A to sample the shopping, the ethnic restaurants, and the entertainment. But they do not stay for long. Although the city is a service centre, the service is distinctly impersonal. Everyone deals with everyone else on the assumption that they are just passing through. Tourists make sure to confirm the price before they buy anything in a shop, and check their change carefully before they leave. The streets are dirty because the tourists drop their litter in a way they would not do at home. Pedestrians have to be careful because cars are driven aggressively and without consideration.

Beyond the theatre district it is mainly gangs of youths that are out on the street at night. Gang membership provides a sense of community that cannot be found elsewhere. Gang loyalty is narrowly focused, however, and inter-gang warfare is common. Older residents stay at home behind locked doors, watching TV dramas about urban crime, interspersed with nostalgic programmes about a vanished rural life. Many old people live alone, having been widowed or divorced and then largely abandoned by their children, who have their own family problems now they are middle-aged. In city A parents cannot even trust their children to honour the implicit contracts of family life.

Although the city is a major financial centre, much of the business is concerned with international trade and investment. Highly reputable international banks deal with highly reputable multinational clients. The major banks do not wish to lend to small local businesses because they are worried about default, and local businessmen do not want to borrow from them anyway because they fear the banks may withdraw funds at short notice in order to gain control of their businesses. Although neither side's suspicions are well founded, the market cannot function because mutual perceptions of risk are so high.

Within the ethnic communities, however, information flows are much stronger because of the regular social interactions within them. Lending risks are lower not only because of better information but because of greater family cohesion—the head of an extended family can pool resources to underwrite an individual member's debts. Ethnic banks have therefore emerged to fill the gap in financing local business.

City B, by contrast, has a large rural hinterland. It developed as a regional market for agricultural produce, and even today many people drive in from neighbouring towns and villages for the banks and shops. The pace of life is much slower than in A. Service in shops is more courteous, traffic moves more slowly, and the city is relatively safe after dark. Tourists increasingly use the city as a base to explore the 'unspoilt' rural areas. There is little for tourists to do in the evenings, but the residents enjoy a lively social life based on church and club meetings. Many are second- or third-generation residents who draw support from other members of their family still living in the area. Community spirit means that there are several well-endowed schools and hospitals too. The main social problem perceived by the adults is that their children watch too much violence on TV.

After many years of economic supremacy, city A is in decline. Stockbrokers retire as soon as they have made sufficient millions to buy a country property near to B, where they spend the money earned in A. Even practising brokers question whether, with further improvements in information technology, it will really be necessary to commute to A. Might they not make better decisions if they were in a lower stress environment like B? If other people think similarly, it may need only a few enterprising migrants to transfer the entire professional information network from A to B.

City B will change as a result, of course. Already some local stores are closing down as national retail chains bid up high-street rents. Catering for the new migrants means that the city is losing some of its distinctive local character. The migrants destroy the very things they come to enjoy.

For B to retain its identity, A must improve the quality of life it offers to potential emigrants. The problem is that, with a low-trust culture, eveyone in A wants to free-ride on other people's efforts. Reversing economic decline requires a co-operative effort which the residents of A are culturally ill equipped to make. Resources that should go into the regeneration of A will be channelled into the expansion of B instead. Rather than conserving B by improving A, low-trust co-ordination mechanisms will leave A to degenerate and B to be spoilt by excessive development.

1.3. The Political Economy of Trust

Recent economic changes, it can be argued, have made trust an even more valuable asset than it was before. In the field of personnel policy, the shortening of product life-cycles, by making innovation a continuous rather than intermittent process, has put training and flexibility at a premium. This has made it very important to encourage loyalty in employees. The continued steady growth of the service sector also means that more people are called to do the kind of intellectual work that is difficult to monitor, or to work on their own, or in very small groups, under conditions where it is uneconomic to employ a supervisor. High service productivity means that workers must be made self-supervising instead.

The greater need for training is putting inter-firm co-operation at a premium too. More sophisticated consumer tastes, and increasingly strict health and safety regulations are increasing the costs of R & D, and so encouraging inter-firm collaboration to avoid duplication in these areas. Higher personal aspirations are also making people increasingly discontented with the inconveniences of city life, and encouraging the decentralization and reagglomeration of economic activity noted above.

Some national cultures seem better placed than others to meet these challenges. The low-trust culture created by the State bureaucracies of the centrally planned economies of East Europe has already precipitated a crisis in these countries. Although much more decentralized, the US economy too has a low-trust culture. This is revealed in the extensive reliance on law, and on the view that competition is the only real safeguard against being cheated over price. Countries such as Sweden and Japan, on the other hand, appear to enjoy a high-trust culture, which has given these highly specialized economies the ability to respond effectively to changing world markets through the flexibility of their highly trained workers.

If this diagnosis is correct, then recent economic trends present a problem for countries that have a low-trust cultural legacy. In many of these countries the dominant political ideology of the 1980s—economic liberalism—has reinforced the attitude of distrust towards fellow-citizens. A rather different ideology, and a very different style of leadership, may be required in the 1990s. The cultural engineering of trust must therefore be high on the policy agenda.

1.4. Microfoundations of Economic Performance

As the preceding examples have shown, culture can influence performance in many different ways. This is not just a practical observation—it is highly relevant at the theoretical level too. Economic theory is essentially a theory of individual decision-making within an interdependent system. Culture can influence both the decision-maker's objectives and his perceptions of the constraints. As a result it can influence all aspects of an individual's behaviour in many different ways (Brubaker 1984; Halperin 1988; Hirsch 1977; Weber 1947). For example, by stimulating a scientific perception of nature, modern Western culture has undermined primitive anthropomorphic attitudes and encouraged the idea that the environment can be controlled. By exhibiting nature as a system it has encouraged the use of a rational calculus in decision-making. This has led, in turn, to the conceptualization of the economy itself as a system, and to the design of bureaucracies as information-handling systems within which decisions can be delegated.

The particular emphasis of this book, however, is on attitudes and beliefs relating to other people. It is concerned with the social rather than the natural environment. The division of labour, together with the spatial dispersion of the population, means that much economic activity is carried out by small groups of people. Some of these groups are fairly permanent—teams of workers in a factory, for example. Conventional economics tends to emphasize relations between groups—between producers and consumers, and workers and employers, for example—rather than relations within groups. A production group within a plant, for example, is usually treated as a 'black box'. In practice, however, the efficiency of co-ordination within groups, as well as between them, is crucial for economic performance. One beneficial effect of the 'supply-side' revolution in economic policy has been a greater emphasis on these microeconomic—as opposed to macroeconomic—determinants of performance.

Other groups are relatively transitory. Examples are traders attending a local market, and people assembling to share the benefits of a public good—listening to the news, attending a performance or ceremony, and so on. The spatial dimension of the economy means that people may also meet inadvertently—this can be a source of problems, for instance when routes to work or to market are congested.

Economic activity over time may, in fact, be conceptualized as a sequence of encounters between individuals (Wolinsky 1987). Culture determines the social environment, and technology the material environment within which these encounters occur. Some of these encounters are one-off chance encounters (trading opportunities, for example) whilst others are planned and recurrent (teamwork, for instance). Overall economic performance may then be related to the success with which the typical encounter fulfils its objectives.

Encounters which are intended to be beneficial can easily turn out to be harmful instead. A trading encounter can lead to harmful results if one of the parties lies about the quality of his supplies or subsequently reneges on some aspect of the agreement. If the encounter is designed to exchange goods on terms agreed at a previous meeting, and one party has already committed most of his costs, the other party may instigate an opportunistic renegotiation of price. Likewise, a production encounter may turn out disappointingly if one of the team members secretly slacks instead of working hard as the others do.

In conventional terminology, it may be said that any encounter between two people potentially affords mutual externalities—each individual's actions impinge on the welfare of the other. These externalities can, in principle, be resolved through contractual arrangements. Since, however, some of the problems relate to behaviour before any contract has been agreed between the parties, the parties alone cannot resolve it. Even after the contract has been agreed, the victim may have difficulty enforcing a penalty against the offender (Sugden 1986).

It is normally necessary, therefore, to invoke a third party with whom each of the parties had a prior contract of some kind. In the case of transactions it is the legal system which typically acts in this capacity. The victim is normally responsible for collecting evidence against the offender, but it is the judiciary that weighs the evidence and enforces the penalty. (Where contraventions of social order are concerned, however, the law itself normally collects the evidence and prosecutes independently of the wishes of the victim.)

In practice, the costs of the legal system are often prohibitive. There are substantial fixed costs in collecting evidence, holding hearings, and determining a judgment, and this means that the legal resolution of minor disputes is inefficient. In practice, the success of encounters in which costs and benefits are small does not hinge on the rule of law but on the existence of goodwill between the parties involved.

1.5. The Nature of Trust

In the absence of a viable legal framework, the crucial question for each individual is whether the other person can be trusted (Gambetta 1988). Can you expect other people with whom no contract exists to forbear from taking advantage—as in a casual encounter when one person has a chance to steal from another? Can you expect people to forbear when negotiating a contract, by giving an honest description of their product and being willing to compromise over price? Can you expect people with whom a contract exists to honour it?

Different social sciences tend to prejudge these issues in different ways. Most economists reject the idea that trust can be a significant factor. Social anthropologists, on the other hand, regard trust as one of the most important factors that bind primitive groups of people together. Marxist sociologists incline to the view that members of the same class (workers, capitalists) can trust each other but members of different classes cannot. In each case the propositions about trust are assumptions which are regarded as self-evident, rather than hypotheses that are put to the test. This means that the crucial question of whether, as a matter of fact, other people can be trusted remains largely unresolved.

An appeal to the conventional concept of economic man suggests that other people will be honest or considerate only when it is expedient for them to be so. This in turn depends upon their preferences and constraints. The former are specific to the person and the latter to their situation. The key aspect of the constraints is the incentive structure. This includes the system of rewards and penalties associated with the current encounter, together with incentives generated by the effect of the outcome of the present encounter on any future encounters in which the parties may be involved.

A conventional economic approach to engineering trust thus depends on manipulating the incentive structure that individuals face. Hostages have an important role in generating suitable incentives. If A holds an asset of great value to B, but of no value to himself, the destruction of the asset provides a powerful way of punishing B if he does not do what A wants. A can therefore trust B, in the sense that A knows that it would be foolish for B to do other than what A wants. If B holds a similar sanction against A then they can be said to trust each other.

In some cases hostages are a natural feature of the situation, whereas in others they must be engineered by a special exchange of assets. In practice one of the most important hostages is an intangible asset, namely reputation. One reason why people with a reputation can be trusted is simply that their reputation is a very valuable asset which they do not wish to lose. Loss of reputation is particularly serious for people who are locked into a group where barriers to exit are high. In fact it can be shown that reputation effects succeed quite well in sustaining trust in small, compact, and isolated social groups.

This kind of trust is no more than enlightened self-interest, of course. The costs of engineering the right incentive structure can be very high. The incentives have to be foolproof, in the sense that no loopholes must be left which the self-interested individual can exploit. The conditions needed to engineer this kind of trust often result in considerable inflexibility—due, for example, to the erection of barriers between groups. It may also require considerable intelligence on the part of group members to calculate that, from the standpoint of the other members, honesty really will be their best policy. To be sure that another self-interested individual can be trusted a large amount of information is needed, and a sophisticated model is required to predict that individual's most expedient strategy (Klein and Leffler 1981).

An alternative mechanism to enlightened self-interest is therefore required. Such an alternative exists, it may be claimed, because of people's capacity for moral commitment. Economic man offers a misleading caricature of human nature because he evaluates only means and does not consider the legitimacy of ends. In practice, individuals have sufficient imagination and empathy to consider the consequences of their actions for others, and to compensate for these by purely internal rewards and punishments. It is perhaps not too fanciful to say that these rewards and punishments are effected by adding to or subtracting from the individual's stock of self-esteem.

Individuals who are morally committed can recognize the potential for moral commitment in others. (A simple theory of human nature is, after all, that 'other people are like me'.) Trust in other people need not, therefore, be based on a complicated calculation of how economic man would respond to a particular incentive structure, but rather on a simple judgement as to whether one is dealing with an 'ethical man' rather than an 'economic man'.

The same conclusion may be arrived at by a slightly different route. Given the prohibitive costs of monitoring small-scale encounters, an obviously superior system is one in which the individuals internalize the external consequences themselves. Each individual, in other words, forbears from actions damaging to the other party. Such activity automatically validates the trust the other party places in him, and helps to sustain mutual trust as a social equilibrium (Parfit 1978; Sen 1977).

The superiority of the moral mechanism over the monitoring system is that it turns to advantage the natural information asymmetry which is the cause of the difficulty in monitoring. The information asymmetry is that people know their own plans, and their own actions, better than others do. Under the moral mechanism, people punish themselves for anti-social behaviour, rather than relying on a third party, such as the legal system, to do it for them. The moral mechanism turns people into self-monitoring agents and so avoids the costs of external monitoring.

A moral sense may be regarded as to some degree innate in most people but, whether one accepts this or not, it is clear that moral sensitivities can be manipulated—for example by parents, role models, priests, and political leaders. It is assumed in this book that each individual is manipulated by a single leader, although it is clear that in practice multiple allegiance is possible. The leader simply personifies the source of moral manipulation. Even where moral influence is clearly dispersed—as with peer-group influence exerted by the many different members of a group—the assumption of a single source of manipulation is retained for the sake of simplicity.

Economies which lack effective moral leadership are obliged to rely very heavily on formal monitoring mechanisms. Economies with strong leadership, by contrast, have a choice of incentive mechanisms. They are not committed to formal monitoring in situations where information is costly to obtain and punishment difficult to enforce. The greater flexibility that effective leadership provides allows overall transaction costs to be reduced, and so overall performance to be improved.

1.6. Leadership

Leadership is a consequence of the division of labour applied to social groups. Most groups have a leader of some kind (McIntosh 1969; McKinley 1965; McPherson 1988). Externally, the leader's role is to represent the group in its relations with other groups—for example, in a primitive group he may act as both statesman and warrior-general. Internally the leader is responsible for co-ordinating group activity. In a group with a formal monitoring system the leader's role is to supervise the system, which usually means acting as a sovereign in upholding the institutions of the law. In a group that does not rely on monitoring, the leader's role is that of a moralist and exemplar who engineers commitment in order to build up trust between the followers. Thus, just as the monitoring of contracts tends to be a specialized function, cultural engineering is specialized too (Dahrendorf 1973).

The engineering of commitment by a leader may well enjoy the active consent of the followers (Lauterbach 1954). There are two main reasons for this. The first is that the leader can answer a couple of crucial questions in the follower's mind. One is 'Are my wants legitimate?' and the other is 'What role should I play within the group?'

Conventional economics, by taking wants as given, entirely overlooks the question of their legitimacy. A philosophical sceptic might argue that such a question is meaningless, or otherwise unanswerable, and ought not to be asked. But, as a matter of fact, it always is asked, as both interview evidence and introspection confirm. And once it has been asked, it is difficult to find peace of mind until it has been answered.

Leaders typically suggest replacing private wants with collective wants, and selfish concerns with altruistic ones. They also try to achieve consensus over collective wants—in other words, preferences are standardized in order to generate a sense of mission within the group. This mission then provides a context within which the question of role can be addressed. Roles which are clearly crucial in fulfilling the mission acquire a higher status than others, and confer non-pecuniary benefits on their holders. The leader can influence the assignment of people to roles by suggesting to particular individuals where their relative strengths and weaknesses (that is, their personal comparative advantages) lie. The moral framework provided by the leader accords social significance to everyday events. The meaning and significance which attach to the patterns and events of everyday life confer on followers a major emotional benefit which supplements the material benefits they derive from group activity.

The second reason why people approve of the leader is that they believe he is influencing other people in the right direction. A typical individual must recognize that he has limited reputation with others, and so little ability to influence them himself. But by supporting a leader of repute he can achieve this influence indirectly. By attaching himself to a leader whose policies he approves of, his support allows the leader to influence other people who then become his followers as well. An individual who understands the leader's role in these terms will be willing to submit to the leader's influence himself, as an example to other followers, provided the other followers are sufficiently numerous that their behaviour is dominant. A key element in the social contract, in other words, is that everyone submits in the same way to the same leader.

1.7. Techniques of Leadership

Unfortunately, leadership today often connotes a style of management which is designed merely to build up people's confidence in the leader himself. This style is particularly notable among military and political leaders, who wish to inspire blind faith in their personal judgement. Such leadership has justifiably acquired a bad reputation. There is, indeed, a strong Western tradition that leaders cannot be trusted. The trust which followers place in their leader gives the leader too much power. 'Power tends to corrupt, and absolute power corrupts absolutely,' it is claimed.

The leadership described in this book has a different orientation, however. The intention is to build up confidence in other people rather than in the leader himself. This in turn reflects the underlying strategy of engineering trust. There are two main aspects of this strategy, and the models in this book explain how these two aspects relate to each other. The first is to make each individual as trustworthy as possible, by building up moral commitment. The second is to make him optimistic about other people's honesty at the same time. Optimism serves two distinct purposes. To begin with, in many situations honesty is more advantageous if it is reciprocated. Secondly, there are other situations where people are so frightened of other people that they are unwilling to participate with them even on a trial basis. Unless people are sufficiently optimistic to give others a chance to prove themselves, they will never learn that they really are trustworthy.

When a leader is dealing with a situation that is essentially short-lived it may not matter if the optimistic beliefs he encourages are unfounded, provided that they have the desired short-run effect. In a more stable long-run situation in which people are repeatedly involved with each other, however, optimism can be sustained only if it turns out to be justified. It is not usually necessary that everyone turns out to be trustworthy in order for widespread honesty to be sustained. There is, however, a critical level of honesty that must be achieved, as an average across the group, if the situation is not to degenerate into universal mutual cheating. The critical proportion of the population that must be honest depends upon the economic incentives involved in encounters within the group. Generally speaking, the more advantageous cheating tends to be at the individual level, the higher is the degree of optimism necessary to combat it. The successful leader needs to know how the incentives govern the critical value, and also what strategies are needed to ensure that this critical value is achieved.

The techniques of leadership vary according to the leader's objective. Where the engineering of trust is concerned, moral rhetoric has an important role. Effective rhetoric must, of course, capture the audience's attention and succeed in holding it for long enough to get the message across. A simple message, vividly illustrated using familiar examples, and delivered with verve, is ideal. But what exactly should the message be?

Moral rhetoric centres on norms of behaviour. It exploits the fact that an individual's mood is strongly affected by whether certain norms have been attained (for a slightly different approach see Loomes and Sugden 1982, 1987). Achieving a high level of output may engender pride and self-satisfaction, whilst a low level of output may induce feelings of inadequacy and shame. What is deemed high or low depends on where the individual sets the norm. By influencing the individual's norm, his emotional response to various levels of output can be manipulated.

Norms can relate to both inputs and outputs. The Protestant work ethic emphasizes norms for the input of effort. This concern with input as well as output is reflected in the more general preoccupation of moral systems with the *intention* as well as the outcome (Marshall 1968). An intentional norm is reflected in a requirement for dedication rather than slacking, and in honest dealing rather than cheating.

The norm of absolute honesty plays a crucial role in this book. It is assumed that people who commit themselves to absolute honesty punish themselves with guilt whenever they cheat. They know that they are committed and can anticipate the guilt they will feel. This guilt is commensurate with material rewards. It enters into the individual's utility function and so forms part of his calculus of choice. It is assumed that while everyone is aware of guilt, people differ in the intensity with which they experience it, according to their particular emotional sensitivity.

In different societies different emotions may be dominant. The role played by guilt in Western societies is, to some extent, played by shame or 'loss of face' in oriental ones. A leader is heavily constrained by the values and customs inherited from the past, and must adapt his message so that it 'resonates' with traditions of the group.

Moral rhetoric can, of course, be subjected to philosophical criticism. Some moral arguments begin with 'self-evident truths' which are difficult, if not impossible, to put to the test. In any case, as Hume emphasized, moral imperatives cannot logically be deduced from factual propositions. Other arguments emphasize the authority conferred by a relevation whose authenticity is difficult to verify. Arguments based upon innate conscience face the objection that conscience may be nurtured by child-rearing instead. The idea that morality is absolute encounters the relativist critique that some moralities clearly endorse actions which are totally repugnant to others.

The analysis in this book is essentially neutral on such issues. It seeks to explain how moral rhetoric is employed rather than to validate or undermine particular arguments. It does, however, provide an instrumental argument in favour of morality, in the sense that moral manipulation is shown to serve material ends. In one sense, of course, this approach turns morality on its head, by justifying the ends as effective means, rather than upholding the ends themselves. In another sense, though, it demonstrates that the morality of honest trading is at least consistent with material ends, and is universalizable, in the sense that if everyone is honest then the economy performs at its best. From this perspective, honest trading may be regarded as a strong contender for the status of a Kantian imperative.

Finally, the kind of reductionist reasoning employed in this book should not be taken to imply that morality is in fact nothing more than a manipulative device that survives because of its economic value. The nature of moral authority is ultimately a philosophical and religious issue, and not an economic one.

1.8. The Culture of the Economics Profession

Because cultural beliefs are rarely articulated explicitly, mutual inconsistencies often go unnoticed. Cross-cultural studies reveal that these beliefs vary dramatically between societies, but people often close their minds to the merits of alternative systems. Alien concepts that cannot be assimilated within their existing system are dismissed as irrelevant. This is much less stressful than reconsidering the system itself. Basic beliefs go unquestioned until a crisis is reached. Only then are inappropriate beliefs discarded and—if the society is fortunate—replaced with more appropriate ones.

Conventional economists exhibit the very kind of cultural bias that they refuse to recognize in others (Klamer 1984; Pen 1985). The culture of an organized profession such as economics can be no less powerful than that of a nation state. Many economists cannot detach themselves from their subject sufficiently to realize that this culture exists. Because this professional culture actually denies that culture affects behaviour, it creates a self-confirming world of delusion. In fact the proposition that culture affects behaviour is never so obvious as in the behaviour of economic professionals themselves (Whitley 1984).

These professional prejudices must be overcome if economics is successfully to handle cultural factors. They are the main reason why, in spite of its technical advantages (Hirschleifer 1985) economics has not contributed more to the analysis of social issues. Political economists—who wish to change the professional culture—have persistently proclaimed a crisis, but little notice has been taken of their claims because their alternatives seem lacking in technical elegance and simplicity. The re-emergence of big issues, however, has heightened the sense of crisis and invested the elaboration of alternatives with a sense of urgency.

The economic turbulence of the 1970s and 1980s, and recent dramatic changes in international relations, has put 'big questions' back on the economist's research agenda. The rise of Japan, for example, and the loss of business confidence within the USA, have rekindled interest in the question of national economic supremacy (Lodge and Vogel 1987; Morishima 1982; Olson 1982).

This question was first systematically addressed by Adam Smith at a time when economics had not become isolated from other social sciences (Barrientos 1988; Campbell 1971; Drakopoulos 1988; Myers 1983). The author of *The Wealth of Nations* was also interested in the nature of moral sentiments and in the growth of scientific knowledge (Smith 1759, 1776, 1795). He was therefore able to accord social and cultural factors a significant place in his explanation of economic growth. A similar emphasis is necessary today if, for example, the Japanese paradox of high economic performance and traditional social structures is to be fully understood. The resolution of issues of this kind requires the wider vision of the social scientist rather than the narrow vision of the typical economic specialist (Barry 1970; Bartlett 1989; Baxter 1988).

1.9. From Economic Man to Ethical Man

The approach in this book is based on the rejection of several of the common prejudices of the narrow economic specialist (though it is not claimed that all these prejudices are held by every specialist). In conventional economics the assumptions made about economic man are carefully tailored to the assumptions made about the environment, and vice versa. People are assumed to exhibit a high degree of rationality because they are also assumed to inhabit a basically very simple world. People are also assumed to be highly autonomous because the only links between them are those mediated by the market system (Casson 1990*a*).

The approach in this book is very different. The assumptions about the environment are altered significantly, and matching changes are made in the assumptions about human behaviour too (Cohen and Axelrod 1984).

1. The idea that individuals are entirely autonomous is rejected. People do not have stable preferences which the economist must take as given (Earl 1983; Felix 1979). Individuals continue to have

a *fixed functional form* for their utility function, but this functional form contains parameters that other people (namely the leader) can set. The retention of a fixed functional form meets the usual objection to the endogeneity of preferences, namely that preferences can simply be altered to provide a *post hoc* rationalization of any kind of behaviour (Becker and Stigler 1977). Specifically, it is assumed that the leader of a group can influence his followers' preferences by controlling these parameters. Followers can thus be manipulated by leaders; only the leaders are autonomous.

2. Utility comprises a material component, analogous to the consumption-related element in conventional theory, and an emotional component which is affected by moral considerations. The emotional component can be influenced by the leader but the material component cannot. The idea that a rational individual will ignore moral considerations is rejected. So too is the idea that morality, when it is relevant, concerns only the economist and the policy-maker, in terms of their formulation of the social-welfare function (Sen 1987). The alternative view presented here is that economies function best when the moral attitudes of the leadership are accepted by the followers and incorporated within their own preferences. 'Ethical man' is a more appropriate model, in this respect, than economic man.

3. The idea that everyone perceives their environment correctly is rejected too (Hargreaves-Heap 1989). Conventional economics tends to assume that the decision-maker's environment is really very simple. In fact most situations are far too complex to handle properly (Loasby 1976). Decision-makers substitute simple models whose variables relate easily to observable aspects of the environment (Hey 1983). The model must perform reasonably well within the usual environment, otherwise it will be discarded, but it may contain errors which cause it to perform badly outside this environment.

An important way of simplifying the perception of the social environment is to identify other individuals with the group to which they belong. This labelling or stereotyping is particularly appropriate when there is a scarcity of information at the individual level. One implication is that reputation adheres to the group rather than the individual. It is assumed in what follows that group reputations are based on a single easily measurable statistic—the average incidence of cheating within the group.

4. The concept of continuity of choice is rejected. In practice individuals often face discrete alternatives, and the margin of substitution is not always active. Optimal decision procedures do not, therefore, imply continuous adjustments to minor variations in the environment, but occasional substantial adjustments triggered at critical points. Another consequence is that, away from the margin, modest fluctuations in the environment are compatible with repetitive behaviour.

5. The emphasis of the analysis is on the general issue of co-ordination within a society, rather than on any one specific manifestation of this. The prejudice of focusing on market trades effected through formal contracts is rejected. Co-ordination is defined simply as Pareto-improvement. It has many manifestations, of which conventional trade is only one. In this book the concept of a trade is replaced with the concept of an encounter, which can also include teamwork, public assembly, chance meetings, and so on. Encounters of certain types have important properties in common, not least of which is that encounters of the *same type* elicit very different behaviour in the one-off and recurrent situations.

Overall, the essence of the approach is to retain the concept of human rationality, but to restrict it merely to a postulate of optimization. The retention of optimization retains the flavour of the methodological individualist approach to economic modelling (Hodgson 1986). It implicitly rejects the traditional sociological view that culture can only be understood from an organic perspective. On the other hand, the postulate that individuals maximize something subject to some constraint does not, when it stands alone, say very much at all. It simply provides a convenient framework within which sets of more specific assumptions can be introduced, and their consequences for behaviour explored in a suitably rigorous way.

There is an important sense, therefore, in which the method of analysis employed in this book is quite conventional, even though the underlying message is not. The similarity with established theory is greatest where the modelling of the leader is concerned. The leader's objectives are taken as autonomous, and the leader is assumed to possess a complete and accurate model of the environment, which the followers do not have. The leader's behaviour is derived through optimization. The leader endogenizes the followers' responses to his manipulation within his calculations. In this respect he behaves as a Stackelberg leader does in oligopoly theory (Friedman 1977).

This optimization approach is applied to a variety of different situations. In Part II of the book the emphasis is on the influence of national leadership on the efficiency of internal trade. The leader engineers a climate of trust between individuals engaged in pairwise encounters, with a special emphasis on trade. Part III is concerned with somewhat larger groupings, such as teams. It explores the conditions under which leadership can be delegated to intermediaries, such as team leaders. Intermediation of trade by brokers and retailers is analysed too. Part IV looks at leadership of collective activity such as the provision of public goods. It is argued that considerations of social justice are an important factor in motivating participation in collective activity. Part V summarizes and synthesizes the analytical results and applies them to the policy issues identified above.

1.10. Relation to Other Theories

It is evident that the approach outlined above is very much aligned with modern institutional economics, with its emphasis on market imperfections caused by transaction costs, and the problem of bounded rationality in devising contractual responses to them (Coase 1937; Hodgson 1988; Williamson 1985). There are, however, two important differences from the usual institutional approach. The first is an explicit retention of optimization—and the consequent rejection of satisficing—which in turn reflects a greater commitment to formal modelling. Bounded rationality is accommodated by substituting a simple optimization model for a more complex one, rather than by abandoning optimization altogether. The second is the specific introduction of the moral and cultural dimension which, although alluded to in transaction cost economics, has not so far occupied centre-stage in the analysis.

There are also affinities with other recent work—notably in non-co-operative game theory (Rasmusen 1989; Smith 1982). One of the intriguing features of contemporary economics is the creative tension between conventional models in the Walrasian tradition, which rely on the law to make contracts costlessly enforceable, and recent game-theoretic models which entirely ignore the law by assuming that contracts are totally unenforceable. Indeed, until recently, applications of non-co-operative game theory were focused instead on international relations—a context in which there is no supranational authority capable of enforcing contracts (treaties).

Rejection of a legal framework also underpins the present approach, but unlike game theory, the matter is not left there. The emergence of a legal system is endogenized, being one possible response to the absence of trust. The legal system is a special type of monitoring system in which people place hostages (their freedom) with an intermediary (the sovereign State) in order to eliminate cheating within the group.

There are even closer affinities with the modern theory of repeated games (Axelrod 1984; Kreps and Wilson 1982). In particular, the so-called trigger strategies, by which cheating is punished, involve elements of both reputation and hostage-taking as discussed in this book. They involve the particular twist that the reputation *is* the hostage (loosely speaking). In this book the emphasis is slightly different, although the same sort of effects appear. It is moral commitment rather than enlightened self-interest that is the basis for reputation, and reputation is mainly significant at the group rather than the individual level.

Finally, there are parallels with recent work by Akerlof (1980, 1983), Jones (1984), and Romer (1984) on social custom and Akerlof (1989), and Akerlof and Yellen (1985) on deviations from rationality, and with research on the free-rider problem in trade-union solidarity (Naylor 1987) and public choice (Schelling 1978 *a*, *b*, 1984). Recent developments in economic psychology are relevant too (Furnham and Lewis 1986; Pieters and van Raaij 1988; Scitovsky 1976). More generally, there are many contexts in which researchers have improvised some kind of moral effect, represented as a deviation from egotistic rationality, in order to explain apparently maverick behaviour (Elster 1979, 1982; Hirschman 1981, 1985; Pemberton 1985; Russell and Thaler 1985; Thaler and Shefrin 1981). The advantage of the present approach is that by facing up squarely to the 'moral dimension' (Etzioni 1988) these *ad hoc* modifications can be subsumed as special cases of a more general systematic approach.

Overall, it is encouraging that many of the analytical building blocks for the present work are already in place. But the architectural plans for the building have not yet been set out. This book can do no more than provide outline plans, but hopefully once these are clear, the remaining details can be filled in.

1.11. Summary

This chapter has outlined a range of issues at the national, local, and corporate levels where cultural factors have an important role. Culture can impact on economic performance in many different ways, but one of the most significant is its influence on trust. Trust reduces transaction costs and so improves the allocation of resources. It is a necessary adjunct to the law in the enforcement of formal contracts, but really comes into its own where informal contracts and understandings are concerned.

Trust can emerge naturally in certain types of situation, but in many cases it needs to be engineered. The social division of labour between leaders and followers provides a mandate for leaders to engineer trust among their followers. The leader makes people trustworthy by emphasizing the moral dimension of behaviour. He stimulates people's emotional responses to their own behaviour. Guilt forms an important element of this response. It is a self-inflicted punishment administered by a self-monitoring individual. It is not sufficient just to make people trustworthy, however. They must be encouraged to place their trust in other people, and this requires that they believe other people are trustworthy too. Otherwise they may regard participation in economic and social activity as too risky. If they do participate, they may anticipate that the material cost of honesty will be too high.

Economists are inclined to dismiss emotion as an irrational factor, but this is methodologically unsound. It merely reflects the cultural bias of their profession. Emotion is simply an element of the preference structure, and can therefore be subjected to conventional optimization analysis.

2

How Leaders Inspire Achievement

2.1. Introduction

This chapter studies the leader as an optimizing agent. It introduces the basic analytical framework which is developed and refined in later chapters. It explains how the leader modifies the incentive structure perceived by a follower simply by associating different emotional penalties and rewards with the follower's various actions.

The framework is designed to generate an interior maximum of the leader's utility. The maximum is 'interior' in the sense that some but not all of the followers will behave as the leader desires. The interior maximum identifies the intensity of manipulation that a rational leader will undertake.

The derivation of the maximum is effected using the familiar economic concepts of marginal benefit and marginal cost. The optimizing intensity of manipulation is the one that equates the marginal benefit of further manipulation to the marginal cost involved. In the context of national leadership, for example, the intensity of manipulation is represented by the level of propaganda and other media activities. The marginal benefit is the incremental improvement in the value of the national product. The marginal cost is the opportunity cost of media services. Equating the two gives the optimal commitment of resources to propaganda activities.

The optimizing strategy relates the intensity of manipulation to the material incentives faced by followers. These incentives can vary from one situation to another. If the problem is one of slacking, for example, then the material incentives refer to the saving of effort the follower can achieve by slacking.

To derive an interior maximum it is necessary that the excess of marginal benefit over marginal cost decline as the intensity of manipulation by the leader increases. The analysis below indicates how this condition is satisfied. It is assumed that the leader has to manipulate a population of people who differ in their sensitivity to manipulation. The marginal benefit attributable to manipulation is related to the number of people who 'get the message' and change their behaviour after an incremental increase in the intensity of manipulation. When sensitivity is uniformly distributed across the population this number declines steadily with respect to the intensity of manipulation. This is because, as the intensity of manipulation increases, a higher proportion of the population has already been converted and those that remain are the least sensitive 'hard cases'. This generates diminishing returns to manipulation, and ensures that manipulation is fixed at a level where some members of the population have not been converted.

The assumption that sensitivity varies across the population has a purely instrumental role in the theory. It is not suggested that it can be directly tested, although it does seem to be fairly plausible. Indirect corroboration is provided by psychological evidence that some people are much more easily aroused than others, for those that are most easily aroused are probably most likely to fall under the leader's influence. There are also numerous anecdotes from practising managers that suggest that some employees are consistently more difficult to motivate than others.

Similarly, the rationality of the leader has a purely instrumental role, as it does throughout conventional economic theory. Because leadership is such a poorly understood phenomenon, however, and instances of apparently ineffective leadership are so common, predictions predicated on the rationality of the leader may prove more unreliable than usual. In applying the theory, therefore, it may be better to relate underperformance to deviations from the optimal strategy rather than to assume maximum performance. In other words, the theory can be interpreted not as a theory of how leaders behave but, alternatively, as a theory of how performance is affected by variations in leadership quality.

2.2. Motivating Achievement

The simplest way to understand leadership is to study an example in which the leader interacts with the followers but the followers do not interact with each other. The most straightforward case is one in which the leader engineers high levels of effort amongst self-employed members of a group. Each member of the group, it is assumed for simplicity, produces the same homogeneous consumption good, and everyone consumes their own output. Each producer is technically independent of the others. This is a reasonable approximation to the engineering of effort among producers by the leader of a nation-state.

In establishing the moral climate the leader has two separate tasks. The first is to establish the norm. By achieving consensus in the group on what constitutes dedication and what slacking, he establishes a common perception of the categories involved. Once these categories are established, each follower perceives a binary choice between dedication (strategy 0) and slacking (strategy 1). All individuals are of the same ability and with dedication can achieve an output $y > 0$. The marginal productivity of effort, Δy $(0 < \Delta y < y)$ is the same for all, and so too is the physiological and psychological cost of effort, $e > 0$.

The second task is to associate guilt with slacking. The guilt penalty is anticipated *ex ante* and—in line with conventional theorizing—the expectations are realized *ex post*. The amount of guilt, $g \geqslant 0$, is commensurate with output and effort, in terms of some *numéraire*.

Individuals exhibit different sensitivities to guilt. Sensitivity captures the extent to which a given intensity of manipulation by the leader impacts upon the follower's feelings of guilt. Sensitivity is denoted s, and is measured on a scale from zero to one (though for mathematical convenience values of s beyond this interval may also be entertained). For each individual the intensity of guilt is the product of the intensity of moral manipulation (which is leader-specific) and the sensitivity factor (which is follower-specific). The intensity of manipulation is measured by the amount of time per period, $\theta \geqslant 0$, that the leader devotes to moral propaganda of one kind or another.

Throughout the book, the convention is used that the information available to any given category of agent (leader, follower, middleman) is displayed in a separate table. Table 2.1 reveals the information available to the representative follower. It is assumed that material and emotional rewards are additive, so that total rewards

TABLE 2.1. *Data for follower's effort choice*

Strategy	Rewards		
	Material	Emotional	Total
Dedication	$y - e$	0	$y - e$
Slacking	$y - \Delta y$	$-g$	$y - \Delta y - g$

are those in the final column. It is further assumed that the rewards are interpersonally comparable, and are fully commensurate with the costs of manipulation considered later.

The table clearly indicates that a follower, faced with a binary choice between dedication and slacking will optimize by choosing dedication if

$$g \geqslant e - \Delta y, \tag{2.1}$$

where

$$g = s\theta. \tag{2.2}$$

The use of a weak inequality in (2.1) indicates the convention that a follower indifferent between dedication and slacking will choose to be dedicated.

The distribution of sensitivity across the population is given by the distribution function $F(s)$,

$$F(0) = 0, \qquad \mathrm{F}(1) = 1.$$

In a group with a finite number of members F is a step function. Although none of the ensuing arguments depends crucially on the continuity of F, it is useful in deriving simple comparative equilibrium results to postulate that F and its first and second derivatives are continuous. For many purposes it is even better to assume that sensitivity is uniformly distributed across an infinite population, so that

$$F(s) = s \qquad (0 \leqslant s \leqslant 1). \tag{2.3}$$

Substituting (2.2) into (2.1) indicates that the critical level of sensitivity at which dedication replaces slacking is

$$s^* = (e - \Delta y)/\theta. \tag{2.4}$$

Let $\mu = 0,1$ indicate the follower's choice of strategy; then the decision rule is

$$\mu = \begin{cases} 0 \text{ if } s \leqslant s^* \\ 1 \text{ if } s > s^*. \end{cases} \tag{2.5}$$

The incidence of slacking is measured by the 'crime rate' q $(0 \leqslant q \leqslant 1)$. Given (2.5), the proportion of followers that will slack is

$$q = F(s^*). \tag{2.6}$$

Thus the crime rate is determined simply by the critical level of sensitivity and the distribution of sensitivity across the group. Note that if $s^* \geqslant 1$ then $q = 1$ (total slacking).

2.3. Policy Objectives

The leader's control problem is to optimize the intensity of manipulation with respect to his objective. But what should his objective be? It is assumed to begin with that the leader is altruistic—the behaviour of a selfish leader is considered in Chapter 11.

An altruistic leader to some extent resembles an economic policy-maker concerned with overall co-ordination of the system. There is, however, a peculiar difficulty because of the endogeneity of followers' preferences. Co-ordination, in the sense of Pareto-improvement, is defined with respect to *given* preferences, while in the present context the problem is solved by preference *change*.

Because material and emotional rewards are, by assumption, additively separable, the problem can be avoided if co-ordination is defined in purely material terms. Two variants of the material approach are worth considering. In the first, called *narrow materialism*, the leader concentrates only on material output and ignores the effort costs involved. This is the approach of many national leaders, who are concerned more with industrial output and per capita GNP than they are with the costs of their followers' efforts. The second measure, called *broad materialism*, includes effort costs. Although it is more satisfactory, it still suffers from the problem of any material measure, which is that it omits the welfare effects of guilt.

A strictly utilitarian approach to welfare would include both the effort penalties experienced by the dedicated and the guilt penalties experienced by those who slack. Such an approach can, however, be criticized on the grounds that utilitarianism is an ethic which may conflict with the ethical system used to identify slacking as wrong in the first place.

Another approach is simply to assert that the leader must work with whatever objective is suggested by his own moral system, irrespective of what his followers feel about it. It is after all, the leader's job to manipulate the followers, rather than the other way round—at least so far as the present model is concerned.

In the light of this discussion, four possible types of leader objective are identified in Table 2.2. Each of the objectives is expressed on a per capita basis—i.e. by dividing total group welfare (as perceived by the leader) by the number of followers involved.

TABLE 2.2. *Objectives of an altruistic leader*

Ethical approach	Leader's objective function	
1. Narrow materialism	$v_1 = y - q\,\Delta y$	(2.7.1)
2. Broad materialism	$v_2 = (y - e) - q\,(\Delta y - e)$	(2.7.2)
3. Utilitarianism	$v_3 = (y - e) - q\,(\Delta y - e + g)$	(2.7.3)
4. Work ethic	$v_4 = e - qe$	(2.7.4)

The specification of the objectives in the materialistic cases (1 and 2) is fairly straightforward, but in the utilitarian case (3) it is necessary to take account of interpersonal differences in the intensity of guilt. Since only those who slack experience guilt, and only the least sensitive slack, the actual guilt experienced, on a per capita basis, is less than it would be if everyone were liable to cheat. This is reflected in the calculation of the average level of guilt,

$$\bar{g} = \theta \int_0^{s^*} s\,dF(s), \tag{2.8}$$

which in the case of a uniform distribution of sensitivity becomes

$$\bar{g} = \begin{cases} 0 & \text{if } \quad s^* < 0 \\ \theta\, s^{*2}/2 & \text{if } 0 \leqslant s^* \leqslant 1, \\ 1 & \text{if } \quad s^* > 1. \end{cases} \tag{2.9}$$

The leader's own morality (case 4) can take many different functional forms. The function illustrated in the table represents an extreme form of the work ethic, in which effort is regarded as good rather than bad, the output it generates is irrelevant, and no sympathy is shown for the guilt experienced by those who slack.

2.4. Nature of the Incentive Problem

The problem of defining co-ordination when preferences are

endogenous is only one of the conceptual problems faced in the present case. There is a second problem which is, to some extent, a consequence of the simplifying assumptions made in the present chapter, and is not so serious in other cases. Because, by assumption, the effort of one follower does not impinge on the productivity of others, there is no externality to resolve in the present case. This means that from a strictly utilitarian view the system might function better without a leader than with one. It is only because the leader's preferences diverge from the utilitarian that, from the leader's point of view, manipulation is required.

In cases (1) and (4) there is an incentive problem only when

$$e > \Delta y \tag{2.10}$$

because only then, when left to their own devices, would followers slack. But from the standpoint of cases (2) and (3) this is exactly what they should do. A broad materialist will optimize with $\theta = 0$ because the cost of effort is so high that dedication is simply not worthwhile. A utilitarian will pursue the same strategy. In this case setting $\theta = 0$ not only avoids too costly effort but eliminates any guilt that would be associated with slacking too.

It is obvious that there is no problem in any case when $e < \Delta y$, for then everyone will spontaneously opt for dedication, which is the best possible outcome. The remaining discussion therefore focuses on a narrowly materialistic leader operating in conditions where (2.10) is satisfied.

2.5. Optimal Manipulation Strategy

The leader is not himself a producer, but he does incur costs—namely costs of manipulation. These costs comprise a fixed cost per follower of making contact (which depends on the number of followers involved, n) and a variable cost per follower which depends on the intensity of manipualtion and is independent of the number of followers involved. The number of followers is treated as a continuous variable, in line with the assumption of a continuous distribution of sensitivity introduced earlier. It is assumed the messages cannot be targeted at specific followers, so that the intensity of manipulation is the same for all followers. Costs, expressed on a per capita basis, are

$$c = \begin{cases} 0 & \text{if } \theta = 0 \\ c_f(n) + c_v\theta & \text{if } \theta > 0 \end{cases} \tag{2.11}$$

where c_f is the fixed cost and c_v the marginal cost.

The leader maximizes a utility which is equal to the value of the objective net of the cost of manipulation. Using a subscript to identify the different cases, in case (1) utility is

$$u_1 = v_1 - \text{c}. \tag{2.12}$$

The leader maximizes u_1 by using θ to influence the crime rate q. In this respect he resembles a Stackelberg leader dealing with a Cournot follower, because he has sufficient information on the followers as a group to know how the crime rate will respond to the intensity of manipulation. Specifically, it is assumed that he knows equations (2.4), (2.6), and the functional form of F. This means, in the case of a uniform distribution of sensitivity, that he correctly perceives the group response function

$$q = \begin{cases} 1 & e - \Delta y > \theta \geqslant 0 \\ (e - \Delta y)/\theta & \theta \geqslant e - \Delta y > 0. \end{cases} \tag{2.13}$$

Substituting (2.11) and (2.13) into (2.12) and taking $\theta > 0$ gives

$$u_1 = y - [\Delta y(e - \Delta y)/\theta] - c_f - c_v\theta.$$

The necessary condition for an interior maximum is

$$du_1/d\theta = \Delta y(e - \Delta y)/\theta^2 - c_v = 0, \tag{2.14}$$

whence

$$\theta^e = [\Delta y(e - \Delta y)/c_v]^{\frac{1}{2}} \tag{2.15.1}$$

$$q^e = \{[(e/\Delta y) - 1]c_v\}^{\frac{1}{2}} \tag{2.15.2}$$

$$v_1^e = y - [\Delta y(e - \Delta y)c_v]^{\frac{1}{2}} \tag{2.15.3}$$

$$u_1^e = y - 2[\Delta y(e - \Delta y)c_v]^{\frac{1}{2}} - c_f. \tag{2.15.4}$$

The intensity of manipulation varies directly with the cost of effort (to overcome material disincentives) and with the marginal productivity of effort (because this makes extra effort more worthwhile so far as the leader is concerned). For obvious reasons it varies inversely with the cost of manipulation. Manipulation is not totally successful in compensating for material disincentives, however, because the crime rate varies directly with the cost of effort too.

The second order condition for a maximum is

$$\mathrm{d}^2u_1/\mathrm{d}\theta^2 = -\Delta y(e - \Delta y)/\theta^3 < 0 \tag{2.16}$$

and this condition is always satisfied given the inequalities specified earlier.

To ascertain whether the local maximum is a global maximum it is necessary to compare u_1^e with $u_1(0)$. The net benefit of manipulation, compared to inaction, is

$$\Delta u_1 = u_1^e - u_1(0),$$

where

$$u_1(0) = y - \Delta y.$$

Manipulation is advantageous if $\Delta u_1 > 0$, i.e. if

$$c_f < \Delta y - 2(\Delta y(e - \Delta y)c_v)^{\frac{1}{2}}. \tag{2.17}$$

The situation is illustrated diagrammatically in Fig. 2.1. The equilibrium is closely analagous to that of a profit-maximizing firm, with the leader's objective playing the role of revenue, and manipulation cost replacing production cost. Intensity of manipulation plays the role of the quantity variable. Fixed costs play exactly the same role in each case. The main difference lies in the existence of a threshold quantity. For $\theta < e - \Delta y$, the marginal revenue product of manipulation is zero because even the most sensitive person does not experience sufficient guilt to make effort worthwhile. Once $\theta \geqslant e - \Delta y$, positive returns are obtained. These positive returns are continuously diminishing. This is because the effectiveness of manipulation depends upon the number of followers moving across the boundary between dedication and slacking. As the intensity of manipulation increases, more and more of the impact is felt by people who have already crossed the boundary, and less and less by the remainder who have still to cross. This is in turn a consequence of the fact that manipulation cannot be targeted just on the marginal followers involved.

The figure has two quadrants. The upper, quadrant A, represents the variation of total revenue and total cost with the intensity of manipulation. The lower quadrant B illustrates the corresponding interplay of marginal revenue and marginal cost.

In quadrant A the expected value of output is indicated by the height of the schedule VWV′. The schedule has a kink at W; to the left of W everyone slacks irrespective of the intensity of manipulation, whereas to the right an increasing proportion of people become dedicated. The segment WV′ rises asymptotically towards the horizontal line YY′, which represents per capita output when everyone is dedicated. Leader's utility is maximized where the vertical discrepancy between VWV′ and CC′ is greatest. The maximum attainable discrepancy is EE′. It corresponds to the point F in quadrant B where the schedule OXX′Z, which is marginal to VWV′, intersects NN′, which is marginal to CC′, from above.

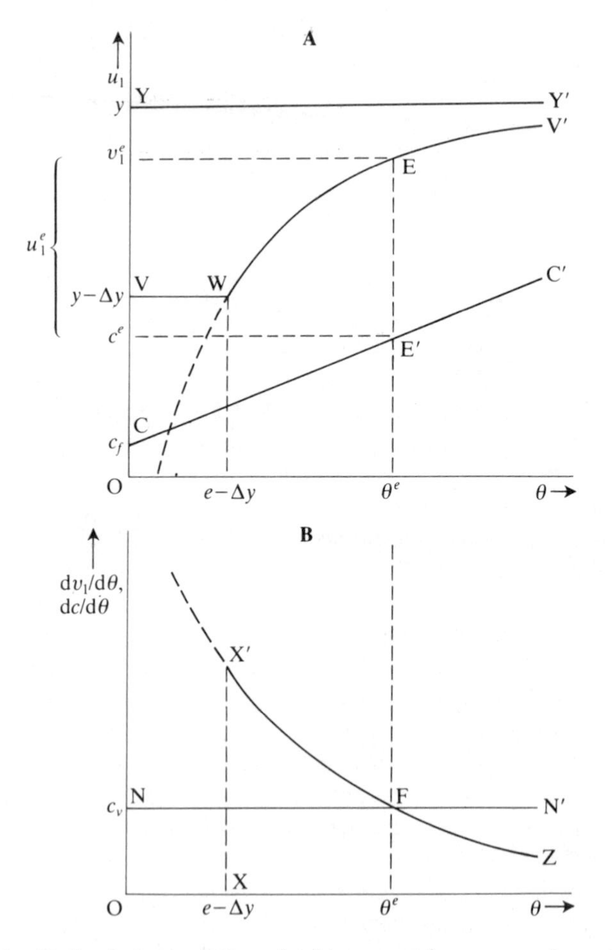

FIG. 2.1 Optimal manipulation of achievement for a narrowly materialistic altruistic leader

2.6 Comparative Statics

The main predictions of the theory may be derived by comparative static analysis. The most important observables are the intensity of manipulation θ^e and per capita output v_1^e. It is also assumed that while individual effort cannot be observed, the crime rate for the group as a whole, q^e, can somehow be estimated. It is also interesting to study the leader utility level u_1^e, even though it is known only to the leader himself.

Changes in these variables are induced by changes in the marginal cost of manipulation, c_v, and by changes in the basic incentive structure, as reflected by the cost of effort, e, and the marginal productivity of effort Δy. The impact coefficients, derived by the differentiation of equations (2.15), are shown in Table 2.3. The easiest way to interpret these results is in terms of the predicted signs shown in Table 2.4.

The negative sign on the first row and first column of Table 2.4 shows, for example, that an increase in the marginal cost of manipulation will reduce the intensity of manipulation, as one would expect. A consequence of a lower intensity of manipulation is a higher crime rate—indicated by a positive sign in the second row of the first column—and by lower output and lower leader utility—as shown in the last two rows of the first column.

The second column shows that a high cost of effort encourages a greater intensity of manipulation, although the intensified manipulation cannot fully compensate for the greater incentive to slack, so that the crime rate is still higher than it would otherwise be. A higher crime rate, in turn, creates lower output and lower leader utility.

Changes in the marginal productivity of effort have two opposing effects on the intensity of manipulation. A high marginal productivity encourages greater manipulation because it means that the penalty exacted by slacking is high. On the other hand, when the cost of effort remains fixed, an increase in the marginal productivity of effort reduces the follower's own incentive to slack and so reduces the need for manipulation. When the cost of effort is less than twice the marginal productivity of effort the second effect dominates the first, and an increase in the marginal productivity of effort reduces manipulation. Conversely, when the cost of effort exceeds twice the marginal productivity of effort, manipulation is increased. In either case, however, the effect on the crime rate is sufficiently beneficial that it falls.

Table 2.3. *Comparative statics: sensitivity of the optimal values (θ^e, q^e, v_1^e, and u_1^e to changes in the parameters c_v, e, Δy)*

	c_v	e	Δy
θ^e	$-\frac{1}{2}[\Delta y(e-\Delta y)/c_v^3]^{\frac{1}{2}}$	$\frac{1}{2}[\Delta y/(e-\Delta y)c_v]^{\frac{1}{2}}$	$[(e/2)-\Delta y]/[c_v\Delta y(e-\Delta y)]^{\frac{1}{2}}$
q^e	$\frac{1}{2}\{[(e/\Delta y)-1]/c_v\}^{\frac{1}{2}}$	$\frac{1}{2}[c_v/\Delta y(e-\Delta y)]^{\frac{1}{2}}$	$-\{c_v/[e/\Delta y)-1]\}^{\frac{1}{2}}\ e/2(\Delta y)^{\frac{1}{2}}$
v_1^e	$-\frac{1}{2}[\Delta y(e-\Delta y)/c_v]^{\frac{1}{2}}$	$-\frac{1}{2}\{c_v/[(e/\Delta y)-1]\}^{\frac{1}{2}}$	$-[(e/2)-\Delta y]\ [c_v/\Delta y(e-\Delta y)]^{\frac{1}{2}}$
u_1^e	$-[\Delta y(e-\Delta y)/c_v]^{\frac{1}{2}}$	$-\{c_v/[(e/\Delta y)-1]\}^{\frac{1}{2}}$	$-(e-2\Delta y]\ [c_v/\Delta y(e-\Delta y)]^{\frac{1}{2}}$

Table 2.4. *Comparative statics: qualitative results*

	c_v	e	Δy	
			$(e<2\Delta y)$	$(e>2\Delta y)$
θ^e	−	+	−	+
q^e	+	+	−	−
v_1^e	−	−	+	−
u_1^e	−	−	+	−

In the first case, where manipulation is reduced as the marginal productivity of effort rises, the significant improvement in followers' own incentives is reflected in higher output and leader utility. By contrast, in the second case, where the more serious consequences of slacking dominate, output and leader utility fall.

To illustrate the application of these results consider the range of factors on which the marginal productivity of effort, the cost of effort, and the cost of manipulation depend. Some relevant factors are listed in the left-hand column of Table 2.5. They may be divided into three broad categories: leader-specific, group-specific, and situation-specific.

TABLE 2.5. *Impact of leader-, group-, and institution-specific factors on the optimal intensity of manipulation*

	Transmission mechanism		
	Cost of manipulation	Cost of effort	Marginal productivity of effort
Leader-specific			
Degree of charisma	+		
Group-specific			
Cost of media services	−		
Average moral sensitivity	+		
Strength of physical constitution		−	
Situation-specific			
Strenuous, unmechanized, and repetitive work		+	
Warm climate		+	
Safety hazards			+

Note: The signs indicate the direction of the impact of the factor on the intensity of manipulation and not on the variable involved in the transmission mechanism.

The most important leader-specific characteristic is the leader's degree of charisma. A high degree of charisma reduces the cost of manipulation and thereby increases the optimal intensity of manipulation. Because the effect is channelled through the cost of manipulation an entry appears in the 'Cost of manipulation' column of the table, and because the effect on manipulation is positive the relevant entry is a plus.

The other main factors influencing the cost of manipulation are the cost of media services and the average moral sensitivity of the group. These are both group-specific factors. High costs of media services discourage manipulation, whilst high moral sensitivity encourages it. Geographically, dispersion is likely to raise media costs because of communications difficulties. Cultural and ethnic homogeneity centred on strong religious traditions and conspicuous social ritual are likely to raise moral sensitivity. Thus both the geography and the cultural traditions of the group are likely to affect the intensity of manipulation.

The cost of effort is another important channel of transmission. This involves a combination of group-specific factors, such as the physical stamina of the typical member, and situation-specific factors relating to the nature of the work and the environment in which it has to be done. Strenuous unmechanized work in a warm climate is particularly costly in physiological terms (the food and water intake required is high). By contrast, mechanized work is normally lighter, and often carried out inside a factory (usually to protect the machines rather than the people), so that the influence of climate may be less. A cold climate may actually favour strenuous work as the physical exercise substitutes for ambient temperature in keeping the body warm. Mechanization sometimes makes work more repetitive, however, and repetition can cause psychological stress—intelligent people find it particularly monotonous. Overall, therefore, mechanization, work organization, and climate can all significantly impact on the intensity of manipulation.

Finally it should be noted that situation specific factors affect not only the cost of effort but its marginal productivity too. The simplest example is where lapses of concentration can endanger safety—a 'Friday afternoon' product may put the customer at serious risk, for example. When slacking is potentially hazardous, the intensity of manipulation needs to be high.

2.7. Sensitivity Analysis

The leader's preferences make a great difference to the leader's optimization strategy. Table 2.6 confirms that, as noted earlier, a leader who is broadly materialistic or utilitarian will not wish to engineer dedication at all. A leader committed to the work ethic,

TABLE 2.6. *A comparison of the optimization strategies associated with the four leadership objectives identified in Table 2.1*

Variables	Cases 1	2, 3	4
θ	$\theta_1^e = [\Delta y(e - \Delta y)/c_v]^{\frac{1}{2}}$	$\theta_2^e = \theta_3^e = 0$	$\theta_4^e = [e(e-\Delta y)/c_v]^{\frac{1}{2}}$
q	$q_1^e = \{[(e/\Delta y)-1]c_v\}^{\frac{1}{2}}$	$q_2^e = q_3^e = 1$	$q_4^e = \{[1-(\Delta y/e)]c_v\}^{\frac{1}{2}}$
v	$v_1^e = y-[\Delta y(e-\Delta y)c_v]^{\frac{1}{2}}$	$v_2^e = v_3^e = y-\Delta y$	$v_4^e = e-[(e-\Delta y)c_v]^{\frac{1}{2}}$
u	$u_1^e = y-2[\Delta y(e-\Delta y)c_v]^{\frac{1}{2}}-c_f$	$u_2^e = u_3^e = y-\Delta y$	$u_4^e = e-2[e(e-\Delta y)c_v]^{\frac{1}{2}}-c_f$

on the other hand, will manipulate more intensively than a narrow materialist, and so achieve the greatest dedication of all. This confirms the intuition that a strong work ethic is more effective in sustaining high output than a narrow materialism.

Given the prevalence of narrow materialism, it is interesting to assess the cost of pursuing this objective from the standpoint of the alternative objectives identified in Table 2.1. Table 2.7 compares the outcome from narrow materialism with the results of optimizing with respect to each of the other objectives instead. From the standpoint of a broad materialist, narrow materialism results in too much manipulation and consequently too much dedication. The same result applies from the utilitarian standpoint too. The welfare cost of pursuing narrow materialism is, however, much greater in the latter case because of the additional guilt induced as well. As already suggested, narrow materialism leads to too little manipulation and consequently too much slacking so far as the work ethic is concerned. Overall, the results suggest that the cost of pursuing narrow materialism can be quite high, although the costs vary a great deal from case to case.

2.8. Differential Abilities

It was assumed above that the leader focuses on an input norm—such as effort—rather than an output norm, although in the present instance this is of no consequence. Any critical level of effort translates into a unique level of output, and so any guilt associated with slacking equates to guilt associated with under-production.

This indifference between input and output norms depends critically, however, on the assumed homogeneity of ability within the population. When abilities differ both the productivity of the dedicated worker and the marginal productivity of effort may be affected. When the former depends on ability but the latter does not the qualitative implications are straightforward. A common output norm will have the greatest impact on those who are just within striking distance of it. If the norm is somewhere just above what a dedicated person of average ability could achieve, then effort will be concentrated amongst those of average ability. Those of high ability will tend to slack because they can achieve the norm without dedication. Those of low ability will also tend to slack, but

Table 2.7. *Distortions created by substituting narrow materialism for alternative leadership objectives*

Intensity of manipulation

$$\theta_1^e - \theta_2^e = [\Delta y\,(e - \Delta y)/c_v]^{\frac{1}{2}} > 0$$
$$\theta_1^e - \theta_3^e = [\Delta y\,(e - \Delta y)/c_v]^{\frac{1}{2}} > 0$$
$$\theta_1^e - \theta_4^e = [1 - (e/\Delta y)]\,[\Delta y\,(e - \Delta y)\,c_v]^{\frac{1}{2}} > \theta$$

Crime rate

$$q_1^e - q_2^e = \{[(e/\Delta y) - 1]c_v\}^{\frac{1}{2}} - 1 < 0$$
$$q_1^e - q_3^e = \{[(e/\Delta y) - 1]c_v\}^{\frac{1}{2}} - 1 < 0$$
$$q_1^e - q_4^e = [1 - (\Delta y/e)]\,[(e/\Delta y) - 1]c_v\}^{\frac{1}{2}} > 0$$

Leader's objective

$$v_2(\theta_1^e) - v_2^e = (\Delta y - e)\,(1 - \{[(e/\Delta y) - 1]c_v\}^{\frac{1}{2}}) < 0$$
$$v_3(\theta_1^e) - v_3^e = (\Delta y - e)\,\{1 + \tfrac{1}{2}\,[(e/\Delta y) - 1]c_v\} - 2(\Delta y - e)\,\{[(e/\Delta y) - 1)]c_v\}^{\frac{1}{2}} - c_f < 0$$
$$v_4(\theta_1^e) - v_4^e = [1 - (e/\Delta y)^{\frac{1}{2}}]\,[e(e - \Delta y)/c_v]^{\frac{1}{2}} < 0$$

Leader's utility

$$u^2(\theta_1^e) - u_2^e = (\Delta y - e) - (2\Delta y - e)\,\{[(e/\Delta y - 1]c_v\}^{\frac{1}{2}} - c_f < 0$$
$$u^3(\theta_1^e) - u_3^e = (\Delta y - e) - \{1 + \tfrac{1}{2}\,[(e/\Delta y) - 1]c_v\} - (2\Delta y - e)\,\{[(e/\Delta y) - 1]c_v\}^{\frac{1}{2}} - c_f < 0$$
$$u^4(\theta_1^e) - u_4^e = [1 - (e/\Delta y)^{\frac{1}{2}}]\,[e(e - \Delta y)c_v] - [1 - (e/\Delta y)]\,[\Delta y(e - \Delta y)c^v]^{\frac{1}{2}} < 0$$

for a different reason—namely that they cannot achieve the norm even with dedication. Only the most sensitive of the low-ability people will work sufficiently hard to achieve the norm—the rest will become demoralized poor performers. Similarly only the most sensitive of the high-ability people will work hard. Thus there will be a concentration of people just above the norm—people of average ability and all degrees of sensitivity, people of low ability but high sensitivity, and people of high ability but low sensitivity. Some distance below this point will be a morality-induced poverty trap in which those who have no chance of achieving the norm make no effort to do so.

The discussion becomes somewhat more complicated if the marginal productivity of effort, as well as the productivity under dedication, varies with ability. A positive association between marginal productivity and ability will tend to widen the dispersion of output, as more low ability workers will become demoralized, whilst the highly sensitive high-ability workers will be able to open up a considerable gap between themselves and the rest of the population. This dispersion will be greater still if there is a positive correlation in the bivariate distribution of ability and sensitivity.

2.9. The Monitoring Alternative

The alternative to manipulation is monitoring. A leader will manipulate only if monitoring is not preferred. There are two aspects to monitoring: the first is the collection of information and the second the application of a material reward or penalty. By making the follower's reward or penalty conditional on the level of effort observed, the leader establishes a material incentive for dedication.

Various incentive systems are possible. The leader could, for example, appropriate the whole of each follower's output and hand back a wage that was related to the level of effort—a wage/bonus system. Alternatively, a lump-sum tax could be applied to all followers, with dedicated followers receiving back a handsome subsidy—a tax/subsidy system. The system considered below is much simpler—namely a fine on slacking.

The fine can be administered in one of two ways. It may be collected in advance and handed back again if performance is satisfactory. This is most appropriate if the leader does not trust the follower and believes that slacking is very likely. The alternative is for the leader to claim the fine only after slacking has been observed. This is appropriate if the leader basically trusts the follower. The leader still needs some mechanism, however, to recover the fine in the unlikely event that slacking occurs. Unless some credible mechanism exists, the incentive mechanism will not work. The leader therefore needs some standing arrangement for the implementation of fines—for example, a police-force and judicial system—to make the incentives credible.

In some cases there is a 'natural' hostage that the leader can take. Thus when the leader intermediates in trade between two followers (as in Section 8.4) the products being exchanged can also function as hostages. If the leader receives both sets of supplies before he passes them on to the other party, then the supply each party expects to receive functions as a hostage. The costs of hostage operation are much lower when natural hostages of this kind are available than when a preliminary exchange of hostages has to be specially arranged.

It is clear, therefore, that a monitoring system must to some degree hold the follower's rewards or penalties hostage—they must be under the leader's control. This in turn gives the leader considerable power over the follower. Thus a police force that can be used to enforce fines can be used for other purposes too. A follower who voluntarily submits to such a system needs to trust the leader not to abuse his power.

In the case of monitoring a leader has a much more direct interest in the choice between input and output norms than under manipulation. This is because his own observation costs may depend critically on what is to be observed. To preserve symmetry with the preceding discussion of guilt, the analysis focuses on the monitoring of inputs although it is probable that in many cases the monitoring of output is cheaper.

Let the level of the fine be $\xi \geqslant 0$. The follower's rewards induced by monitoring are illustrated in Table 2.8. It is a trivial result that all followers will be dedicated if

$$\xi \geqslant e - \Delta y.$$

TABLE 2.8. *Follower's data set under monitoring*

Strategy	Normal material rewards	Fine	Total material rewards
Dedication	$y - e$	0	$y - e$
Slacking	$y - \Delta y$	$-\xi$	$y - \Delta y - \xi$

Everyone is affected because, by assumption, everyone has the same sensitivity to material incentives. The issue of diminishing marginal returns to the leader's intervention does not therefore arise.

It is assumed that monitoring costs are fixed costs entirely independent of the size of the penalty involved. Using a prime to signify variables pertaining to monitoring, we have

$$c' = \begin{cases} 0 & \text{if } \xi = 0 \\ c_f'(n) & \text{if } \xi > 0. \end{cases} \tag{2.19}$$

Since the marginal cost of increasing the size of hostage is zero, the optimal monitoring strategy is to set the hostage value a little in excess of the critical value, thereby generating total dedication and the highest possible material reward:

$$q'^e = 0 \tag{2.20.1}$$

$$v_1'^e = y \tag{2.20.2}$$

$$u_1'^e = y - c_f'. \tag{2.20.3}$$

Comparing monitoring with inactivity, monitoring is preferred if

$$c_f' < \Delta y. \tag{2.21}$$

2.10. Relative Transaction Costs

To compare manipulation and monitoring it is useful to reformulate the leader's problem in terms of transaction costs. The leader's relation with a follower may be regarded as a transaction in which the follower agrees (informally) to work hard in return for some benefit—emotional or material—conferred by the leader. Equivalently, the leader–follower transaction may be perceived in terms of agency. In the present context, agency costs are essentially transaction costs perceived from the leader's point of view. The agency problem confronted by the leader is the problem of getting the follower to be as dedicated as the leader wishes him to be. When reformulated in this way, the maximization of leaders' utility become equivalent to the minimization of transaction (or agency) costs. The choice between monitoring and manipulation involves selecting the leadership technique that affords minimum transaction cost.

The transaction cost of monitoring is simply the fixed cost mentioned earlier:

$$t' = c_f'. \tag{2.22}$$

The transaction cost of manipulation has two components, however. The first is the direct cost incurred by the leader, and the second the indirect cost attributed to underperformance by slackers:

$$t = t_1 + t_2$$

where

$$t_1 = c$$

$$t_2 = y - v_1.$$

Thus the total transaction costs of manipulation are

$$t = y - v_1 - c = y - u_1. \tag{2.23}$$

Manipulation is chosen if

$$t \leqslant t'.$$

Using (2.15.4) and (2.22), this condition becomes

$$c_f < c_f' - 2[\Delta y\ (e - \Delta y)c_v]^{\frac{1}{2}}. \qquad (2.24)$$

The situation is illustrated in Fig. 2.2. The fixed cost of monitoring is indicated by the height of the horizontal line MM′. The direct cost of manipulation is illustrated by the schedule CC′ transcribed from Fig. 2.1. The indirect cost of manipulation IJI′ measures the output loss associated with slacking, and is equal to the vertical discrepancy betweeen schedules YY′ and VWV′ in Fig. 2.1.

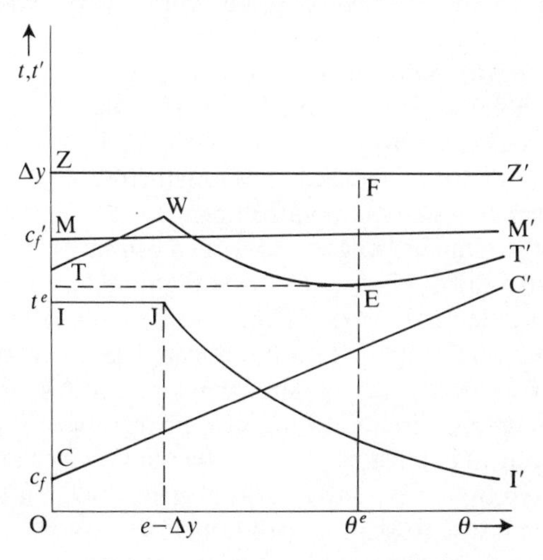

FIG. 2.2 Choice between monitoring and manipulation

As shown, the fixed cost of monitoring c_f' exceeds the minimum total cost incurred by manipulation t^e, so that manipulation is preferred. The optimal intensity of manipulation is, of course, θ^e. At this intensity of manipulation total transaction costs are less than the gains from manipulation, Δy, so that the transaction between leader and follower is, on balance, profitable from the leader's point of view. The profitability of manipulation is indicated by the fact that the height of the transaction gain schedule ZZ′ at F exceeds the height E of the transaction cost schedule TWT′.

Taking up the earlier examples in Section 2.6 it is evident that monitoring is particularly attractive to leaders who lack charisma and are faced with high media costs. Monitoring is also particularly suitable for strenuous outdoor work in warm climates, where manipulation must be quite intense to have much effect. It is often alleged that the management of work is extremely harsh in industries involving strenuous outdoor work, such as construction, and in countries with warm climates (where it may be associated with large public works under despotic political regimes). The analysis above provides some theoretical basis for such allegations.

Perhaps the most important and fruitful comparisons, however, can be drawn from situations in which the monitoring costs differ. If the workers are principally engaged on intellectual activity, for example, then the monitoring of inputs will be very expensive. If the workers are engaged mainly on manual activity, on the other hand, then monitoring may be quite cheap, particularly if they are agglomerated together in a factory. The theory therefore predicts, in line with the speculations advanced in Chapter 1, that intellectual workers will tend to be motivated by high-trust leadership and manual workers by low-trust monitoring.

This line of argument suggests that in most cases the leader's charisma must be of a kind that affects highly educated workers engaged in intellectual work. The so-called charisma of the demagogue, who appeals to poorly educated people, may have little value in a society where large agglomerations of manual workers are best co-ordinated by monitoring instead.

It also suggests that a distinctively intellectual style of leadership may be required by societies that wish to specialize in intellectual work. Extending low-trust monitoring mechanisms from manual work to intellectual work will pose serious problems for societies wishing to specialize more in intellectual areas such as professional services and R & D. Similarly borrowing leadership techniques from areas where the motivation of manual work is paramount (for example, a conscripted army) will be ineffective too. Service economies based on high intellectual productivity require the evolution of a quite distinctive leadership style, which traditional political and military models of leadership cannot provide.

The point should not be exaggerated, however. Not all intellectual activity benefits from leadership in place of monitoring. Some intellectual slacking can have extremely hazardous consequences. The safety of medical practices is a case in point. In such cases the difficulty of manipulating the least sensitive individuals encourages monitoring instead.

A more comprehensive analysis of the choice between leadership and monitoring would require the relaxation of two key assumptions made above. The first is that sensitivity of *material* incentives is entirely uniform and only sensitivity to *emotional* incentives varies. It is this assumption that allows monitoring to perform much better than manipulation in respect of the total rather than partial elimination of slacking in the situations discussed above.

The second is the view that leadership and monitoring are strict alternatives. The main reason for regarding them as alternatives is that it is difficult for a leader to demand spontaneous integrity from his followers when it is clear from his monitoring practices that he does not really trust them. In cases where safety is threatened by slacking, however, it is not difficult to justify monitoring on the basis of policing a minority of followers who are not representative of the group as a whole. A more sophisticated analysis would recognize that leadership and monitoring can complement each other, and that investing in both may well be worthwhile in hazardous operations where the marginal productivity of effort is high.

2.11. Summary

This chapter has established that the concept of moral manipulation by a leader can be placed on a rigorous footing. The leader has been modelled as a Stackelberg leader who manipulates a group of followers by changing their preferences. Preference change is effected by making followers' utility depend not only on material rewards but also on the magnitude of a guilt penalty. The leader controls the group by setting the intensity of manipulation. The flow of rhetoric impacts differentially on individual guilt, according to personal sensitivity.

The focus in this chapter has been on the incidence of slacking amongst technologically independent workers. Technological independence means that there are no externalities between different members of the group. Externalities will be introduced in the next chapter. The average incidence of slacking is measured by the crime rate for the group. Increasing the intensity of manipulation reduces the crime rate as it improves the overall performance of the group.

What is meant by improvement depends, however, on the exact

nature of the leader's preferences. Of the four main preference systems discussed, one—narrow materialism—has been selected as the basis for subsequent analysis. A narrowly materialistic leader disregards both the physiological costs of effort incurred by his followers and the guilt he induces in those who slack. The leader is simply concerned with maximizing the sum of the material rewards of the members of the group.

The optimal strategy of manipulation is determined by the interplay of leader-specific, group-specific, and situation-specific factors. These impact on the intensity of manipulation through the cost of manipulation, the cost of effort, and the marginal productivity of effort. The cost of manipulation is partly leader-specific, since a highly charismatic leader will face lower costs, and partly situation-specific, as reflected in the opportunity cost of media services. The cost of effort is partly group-specific, reflecting the exact conditions under which they have to work. The marginal productivity of effort is largely situation-specific, as reflected, for example, in the safety hazards that slacking might cause.

The leader's choice of manipulation strategy is actually just one aspect of a wider choice between manipulation on the one hand and monitoring on the other. It has been assumed that monitoring is completely effective in identifying slackers, and that slacking is punished by a material fine to which all followers are equally sensitive.

A leader facing this wider choice will tend to opt for monitoring in situations where the most intensive manipulation would otherwise be necessary. Strict monitoring is thus a common response to hazardous situations. Monitoring is also preferred where manipulation is costly—where media services are expensive, for example, or the leader is lacking in charisma.

Monitoring is least appropriate where monitoring costs are relatively high. This apparently trivial proposition has important implications because monitoring costs do, in fact, vary a great deal between one situation and another. Manual labour, for example, typically incurs low monitoring costs, and therefore offers limited scope for leadership through manipulation. Intellectual work, on the other hand, encounters higher monitoring costs, and so in this context manipulation is more appropriate. The style of leadership required where intellectual work is concerned is, however, very different from that of the demagogue who has so often played the role of political or military leader in the past.

PART II

Co-ordination with Pairwise Encounters

3

How Culture Sustains Trade

3.1. Introduction

This chapter uses the analytical techniques expounded in Chapter 2 to address the central issue identified in Chapter 1—namely the problem of co-ordinating encounters in which mutual externalities are involved. An encounter between two people is represented as a 2×2 non-co-operative game. Each individual can choose to be honest or to cheat. 'Honesty' and 'cheating' are the analogues of 'dedication' and 'slacking' in the previous chapter. Only mutual honesty achieves full co-ordination.

In the conventional theory of games, encounters are classified by the nature of the material incentives involved. In some games, such as the Prisoners' Dilemma (PD) it always pays to cheat. Trade based upon exchange of contracts provides an example of this. Each party normally stands to gain by reneging on a contract, independently of whether his partner honours it. If the other party is honest it is advantageous to take his goods and give nothing in return. If the other party cheats, it is advantageous not to have given him any goods in the first place. When a one-off trade is concerned, therefore, honesty is never the best policy.

In another type of game, sometimes called an Assurance game, it only pays to be honest if the other party is honest too. Consider, for example, a two-person team that is only as strong as its weakest link. Associate co-operation with honesty and selfishness with default; then honesty is worthwhile only if the other team member is honest too. It takes only one of the team members to be dishonest and the whole endeavour falls through. If the other party is honest, however, then it is self-defeating to cheat so long as the anticipated share of the team product exceeds the cost of co-operation involved.

The principle of Assurance can be characterized as 'I will if you will; I won't if you won't.' The opposite of Assurance is the game of Chicken, in which the best strategy for any player is 'I will if you won't; I won't if you will.' Consider, for example, two motorists who are approaching a narrow bridge from opposite directions. One of them must give way if a collision is to be avoided. Each prefers that the other gives way, but would rather give way than be involved in a collision. It is better that both give way (and then toss a coin to decide who goes first) than that neither of them does.

Another example of Chicken is the price negotiation that precedes an exchange of contracts. Each party would prefer the other to take a soft line in negotiations, so they can take a hard line and extract a major concession over price. Both would rather concede, however, than have the trade fall through. Mutual concession—such as a 'split the difference' compromise—is mutually advantageous compared to no trade at all.

The most desirable kind of game is Harmony, where both parties prefer to be honest whether or not the other party cheats. There are few encounters which are naturally harmonious when only the material incentives are considered. Manipulation of players' preference by a leader can, however, transform a game from one kind into another. This provides a mechanism for transforming games such as Prisoner's Dilemma, Chicken, and Assurance into a game of Harmony instead.

In the case of trade, for example, the leader may associate guilt with default. By associating guilt with the intention to cheat, honesty will be induced whether the partner is expected to be honest or not. If the guilt is sufficiently intense then it always pays to be honest, and Harmony is the result.

To emphasize the generality of this effect, consider a rather different example in which the PD arises outside the market environment. The issue is the quality of the environment—an issue that commonly arises when markets are missing or the prices of certain property rights are arbitrarily constrained to zero. The problem concerns noise levels when there is no tradable 'right to peace and quiet'. Two neighbours enjoy listening to very different kinds of music, and each positively dislikes the other's taste. Each can hear what the other is playing. They would prefer their neighbour to be silent so they can listen to their own kind of music in peace. But if their neighbour plays music they would still rather play their own music simply to drown out the neighbour's noise. When choosing between complete silence and cacophony, however, silence is preferred. But because playing music is always individually rational, cacophony invariably occurs.

By establishing a code that disturbing neighbours is morally wrong, the missing property rights to peace and quiet can be effectively replaced. The leader associates guilt with noise-making, and so encourages silence instead. Both neighbours opt for silence, and the cacophony is eliminated making both better off than before.

This chapter examines both the general case, in which an encounter can be one of any of the types discussed above, and the special case of trade, where the encounter is a PD. The general model is set up in Section 3.3. The special case of trade is examined in Sections 3.4 and 3.5, which show how trade naturally generalizes some of the results obtained in Chapter 2. The general case is analysed in Sections 3.6 and 3.7 and a number of complicating factors are discussed. The conclusions are summarized in Section 3.8.

3.2. Modelling Encounters

In conventional game theory it is assumed that each player knows not only his own but also his partner's rewards. This assumption is here replaced by the weaker and more plausible assumption that each player knows only his own rewards. Players operate with an incomplete model of their environment. This incompleteness reflects their bounded rationality and the complexity of the social environment that they face.

Bounded rationality is sometimes accommodated by introducing information costs, assuming 'satisficing' rather than maximizing behaviour or postulating a simple behavioural rule (Simon 1983). In the present context, however, it simply means that individuals use an oversimplified maximizing model in which, instead of calculating their partner's optimal strategy, they associate a subjective probability with their partner's decision to cheat.

In the one-off encounters studied in this chapter the subjective probability is something that the leader may be able to influence. Two cases are in fact considered; in the first, subjective beliefs are distributed across the population in a manner which the leader cannot control, while in the second the leader can standardize beliefs through a suitable announcement strategy. It is usually best for the leader to standardize the group on optimism—though aiming for complete optimism may not always be the best policy. A surprising qualification is that in some cases standardizing on pessimism is more appropriate.

Another difference from conventional game theory is that the emphasis is on a group of players rather than just a single pair. Pairs are formed by matching individuals within the group. Nevertheless, one important aspect of conventional game theory is retained. Participation in the game is compulsory—players cannot opt out. This assumption is, in fact, a serious limitation where the analysis of group behaviour is concerned, and is relaxed later on in Chapter 5.

Consider, therefore, a large group of fixed size, n, within which pairwise encounters occur. The group is the basic economic unit in this analysis and the co-ordination of encounters governs group performance. In a material sense, everyone is identical—they experience the same material rewards from the same pairwise combination of strategies. In the emotional dimension, however, people differ in their sensitivities, along the lines described in Chapter 2.

In each period each individual experiences one encounter with a randomly selected partner. This chapter concentrates on encounters over just a single period: recurrent encounters are considered in Chapter 4. At the beginning of the period there is no accumulated experience. In the absence of any announcement by the leader, everyone forms their own opinion of whether their partner is likely to be honest or not.

The concept of an encounter is a quite general one, and embraces a number of special cases. Trade is particularly interesting, not only because of its immense practical importance, but because it provides a convenient transition between the model of independent production developed in Chapter 2 and the model of team production developed in Chapter 7. Trade involves two technically independent but contractually linked activities—namely the transfer of a good (or service) from one individual to another, and a matching transfer in the other direction. Trade therefore highlights the opportunities for default that arise for purely contractual reasons, quite independently of the technological externalities associated with team production.

It should also be noted that the monitoring of inputs is again equivalents to the monitoring of outputs, just as it was in Chapter 2. If problems of 'loss in transit' are ignored, then the only reason why the recipient fails to get the goods he was promised is because his trading partner has defaulted. Monitoring despatch (the 'input' end) and delivery (the 'output' end) are therefore equivalent. By contrast, in team production the monitoring of team output does not permit inferences about individual inputs (although the converse is true if the technology is completely known).

3.3. The Incentive to Cheat

The material rewards generated by an encounter are indicated in Table 3.1. Mutual honesty rewards each person with $h \geqslant 0$. Encounters can, however, make people worse off rather than better off; with mutual cheating it would sometimes have been better that the encounter had never occurred at all. The *symmetric co-ordination gain* $a > 0$ measures the gain from mutual honesty compared to mutual cheating, and may well exceed the basic gain h.

TABLE 3.1. *Material rewards from an encounter*

Follower's strategy	Partner's strategy	
	Honesty	Cheating
Honesty	h	$h - b - d$
Cheating	$h + b$	$h - a$

The *incentive to cheat* when the other party is honest is b. Now if cheating merely redistributes rewards then what the cheater gains the honest victim loses; thus the cheat who gains $h + b$ rather than h causes the honest victim to gain only $h - b$ instead of h, leaving the total gain to the pair of them unchanged at $2h$. This case in which cheating merely redistributes rewards is ruled out by assuming that the victim loses a further amount $d > 0$. This is the *asymmetric co-ordination gain*—it measures the gain to mutual honesty compared to one person taking the other for a sucker.

The structure of emotional rewards is illustrated in Table 3.2. The important point to note is that guilt is associated both with

TABLE 3.2. *Emotional rewards from an encounter*

Follower's strategy	Partner's strategy	
	Honesty	Cheating
Honesty	0	0
Cheating	g	g

cheating an honest person and with cheating on a cheat. This assumption is relaxed in Chapter 6—first to allow for the possibility that guilt is associated just with cheating an honest victim, and then to allow for the pleasure of vengeance against those who cheat.

The present approach may be justified on two grounds. The first is philosophical: guilt is related to moral intent, and the intent to cheat is formed before it is known whether the other party cheats as well. The second is more practical: confining guilt to cheating the innocent introduces non-linearities which complicate the analysis.

Combining the material and emotional rewards within the simplified follower's decision model outlined above gives the data set shown in Table 3.3. The follower associates a subjective probability p $(0 \leqslant p \leqslant 1)$ with the possibility that his partner cheats. The follower prefers honesty to cheating if the expected value of the benefit, as he perceives it, is non-negative:

$$(1 - p)(g - b) + p\,(g + a - b - d) \geqslant 0,$$

i.e.

$$g \geqslant b + kp \tag{3.1}$$

where

$$k = d - a \tag{3.2}$$

TABLE 3.3. *Follower's data set for an encounter*

Follower's strategy	Partner's strategy	
	Honesty	Cheating
Honesty	h	$h - b - d$
Cheating	$h + b - g$	$h - a - g$
Perceived probability	$1 - p$	p

measures the excess of the incentive to cheat a cheat over the incentive to cheat an honest partner. The larger is d, the more you lose by being taken for a sucker, compared to what you would gain by cheating someone honest. The smaller is a, the less you lose by cheating a cheat compared to reciprocating honesty to an honest party.

For an individual of sensitivity s ($0 \leqslant s \leqslant 1$) facing an intensity of manipulation θ, the guilt (compare (2.2)) is

$$g = s\theta, \tag{3.3}$$

whence, from (3.1) and (3.3), the critical level of sensitivity which induces honesty is

$$s^* = (b + kp)/\theta. \tag{3.4}$$

Let $F(s,p)$ represent the bivariate distribution, across the group, of sensitivity s and probability p. The incentive to default across the group is captured by the crime rate. The crime rate q corresponds to the proportion of the population which combines a probability p with a sensitivity $s < s^*$. Thus, in general, q depends on the parameters of F, the values b and k, and the intensity of manipulation, θ.

In the special case $k = 0$, s^* is independent of p,

$$s^* = b/\theta, \tag{3.5}$$

and

$$q = F(s^*, 1). \tag{3.6}$$

In other cases the derivation of q is more complex, as indicated in Section 3.5.

An altruistic leader pursuing narrow materialism maximizes

$$u = v - c, \tag{3.7}$$

where v is the expected value of the material reward of a representative follower and c is the cost of manipulation (v corresponds to the reward v_1 defined in Chapter 2). With a crime rate q the probability that both are honest is $(1 - q)^2$ and in this case both receive a reward h. The probability that a cheat takes advantage of an honest victim is $2q(1 - q)$, and the reward, averaged over cheat and victim, is $h - (d/2)$. Finally, the probability that both cheat is q^2, and in this case each receives a reward $h - a$. Thus the expected value of the material reward is

$$\begin{aligned} v &= (1 - q)^2 h + q(1 - q)(2h - d) + q^2(h - a) \\ &= h - dq + kq^2. \end{aligned}$$

For $k = 0$, v is a linear function of q, but otherwise it is quadratic. To eliminate the non-linearity in the order to permit a simple solution of the leader's optimization problem, a first-order approximation is taken about the mid-point of the unit interval over which q is defined. Using the first two terms of a Taylor's series expansion about $q = \frac{1}{2}$ gives

$$v = v_0 - aq, \tag{3.8}$$

where

$$v_0 = h - (k/4). \tag{3.9}$$

The approximation is illustrated in Fig. 3.1.

Note: The case illustrated is for $k > 0$.

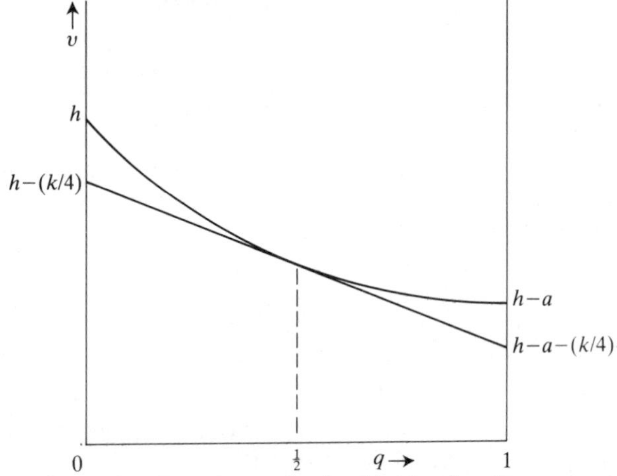

FIG. 3.1 Approximation of a quadratic objective function

As before, the cost of manipulation is

$$c = \begin{cases} 0 & \theta = 0 \\ c_f + c_v\theta & \theta > 0. \end{cases} \tag{3.10}$$

Given the leader's knowledge of the follower's decision data, and the parameters of the distribution F, he can predict the crime rate as a function of the intensity of manipulation. The leader's utility can be then be expressed as a function of θ and maximized accordingly.

3.4. A Trading Economy

Consider a symmetric trading situation in which each party exchanges a single unit of a good he values at b for a single unit of a good he values at $a + b$ $(a, b > 0)$. Supplies have to be despatched before purchases are received, and the individual does not know whether the purchases are on their way at the time he commits himself to despatch. This generates the structure of rewards shown in Table 3.4, which corresponds to Table 3.3 with $h = d = a$, which in turn means that $k = 0$.

TABLE 3.4 *Follower's data set for a trading encounter*

Follower's strategy	Partner's strategy	
	Honesty	Cheating
Honesty	a	$-b$
Cheating	$a + b - g$	$-g$
Perceived probability	$1 - p$	p

In the trade situation, therefore, the critical level of guilt needed to guarantee honest trade is independent of the perceived probability that the other party will cheat. This is because of the technical separability of the two parts of the exchange. The gain to cheating is that one party keeps the product that the other party is expecting to receive, and this product is of value b independently of whether the other part of the transaction is fulfilled.

The role of manipulation is to remove the PD by associating an emotional penalty with cheating. Since the material gain to cheating is b, and is the same for everyone, manipulation will have no effect until the intensity reaches b. Only then will the most sensitive decide not to cheat. As the intensity is increased beyond b, others will decide to be honest too, but, given a suitable distribution F, the number crossing the threshold to honesty will decline steadily as those that remain as cheats are increasingly insensitive to guilt. Eventually some finite optimal intensity of manipulation is reached at which the marginal cost of greater intensity is just equal to the expected value of the marginal gain from greater honesty that will result. The only qualification is that the fixed cost of manipulation should not be so high that manipulation of any kind is uneconomic.

The best way to illustrate the optimization is to postulate a uniform bivariate distribution of sensitivity and probability

$$F(s, p) = sp \qquad (0 \leqslant s, p \leqslant 1). \tag{3.11}$$

Equation (3.11) implies that members of the group are uniformly distributed over the unit square in $s - p$ space, with perceived probability being independent of the sensitivity of the individual concerned.

Because $k = 0$ the derivation of the crime rate is very straightforward. Since (3.5) and (3.6) apply,

$$q = F(s^*, 1) = F(b/\theta, 1) = \begin{cases} 1 & \theta < b \\ b/\theta & \theta \geqslant b. \end{cases} \tag{3.12}$$

Substituting (3.8)–(3.10) and (3.12) into (3.7) and taking $\theta \geqslant b$ gives

$$u = a - c_f - (ab/\theta)\, c_v \theta. \tag{3.13}$$

The first-order condition for an interior maximum

$$\mathrm{d}u/\mathrm{d}\theta = (ab/\theta^2) - c_v = 0 \tag{3.14}$$

gives

$$\theta^e = (ab/c_v)^{\frac{1}{2}} \tag{3.15.1}$$

$$q^e = (bc_v/a)^{\frac{1}{2}} \tag{3.15.2}$$

$$v^e = v_0 - (abc_v)^{\frac{1}{2}} \tag{3.15.3}$$

$$u^e = v_0 - 2(abc_v)^{\frac{1}{2}} - c_f. \tag{3.15.4}$$

The second-order condition for a maximum is

$$\mathrm{d}^2u/\mathrm{d}\theta^2 = -2ab/\theta^3 < 0, \tag{3.16}$$

and this condition is always satisfied given the inequalities specified earlier. The net benefit of manipulation compared with inaction is

$$\Delta u = u^e - u(0),$$

where

$$u(0) = v_0 - a.$$

Manipulation is advantageous if $\Delta u > 0$, i.e. if

$$c_f < a - 2(abc_v)^{\frac{1}{2}}. \tag{3.17}$$

It can be seen that equations (3.15)–(3.17) are the exact analogue of equations (2.15)–(2.17), with

$$a = \Delta y, \qquad b = e - \Delta y \tag{3.18}$$

being the symmetric co-ordination gain and the incentive to cheat in each case. The similarity is quite remarkable given that the trade model involves strategic interactions between followers, whereas the achievement model of Chapter 2 does not. The explanation, of course, is that since, in the trade model, $k = 0$, subjective beliefs do not influence the incentive to cheat, so it is only th sensitivity to guilt, and not the probability perceptions, that governs follower behaviour—just as in the achievement case.

There is, however, an important difference between the two cases so far as the welfare implications are concerned. The objectives of broad materialism and narrow materialism coincide in the trade case, because co-ordination involves no effort penalty for the follower and so the utility gain Δu is a more acceptable measure of welfare gain than the utility gain Δu_1 of Chapter 2. In the trade case, as in the achievement case, however, a utilitarian would regard the level of manipulation as excessive, because of the guilt induced in the cheats.

In the trade context, the analogue of the work ethic is the honesty ethic—the pursuit of honest trading practices, not as a means to material co-ordination, but as an end in themselves. With this modification, the analogy with Chapter 2 still holds—the honesty ethic suggests that by following materialist objectives the intensity of manipulation will be set too low. (This result applies, however, only if the weight attached to honesty exceeds the material gain to co-ordination, a.)

When interpreting the results (3.15), the obvious point to begin with is that the optimal intensity of manipulation is homogeneous of degree one in a and b; this means that as the value of the product traded increases, the intensity of manipulation increases in direct proportion. On the other hand, the crime rate is homogeneous of degree zero in a and b, so as the value of the product and the associated intensity of manipulation increase in line with each other, the crime rate remains unchanged.

When the cost of manipulation c_v is considered as well, the homogeneity of the intensity of manipulation falls to one half; this implies that if the cost of manipulation increases in line with the value of the product then the intensity of manipulation will still rise to some extent, but even so the crime rate will increase as well. This is because the increase in the value of the product increases the material incentive to cheat at a time when the rising cost of manipulation makes it difficult to achieve a corresponding increase in the emotional penalties involved.

The second point to note is that the crime rate is related to the ratio of the supplier's valuation of the product, b, to the gain from trade a. The crime rate is higher, the higher is the proportion of the buyers' valuation that is matched by the seller's valuation. This proportion is likely to be high for versatile goods such as land, but much lower for highly customized goods such as professional services, where the value to the customer is much greater than the opportunity cost of the professional's time. This suggests that moral manipulation will be more effective with customized goods such as professional services than with versatile goods such as land.

The third point concerns the effects of variations in the cost of manipulation c_v. When the traders come from very different ethnic or religious groups, or are highly dispersed over space, suitable moral rhetoric and effective communication may be difficult to achieve. Thus in international trade, for example, moral manipulation may be much less viable than in internal national trade.

3.5. Comparison with Monitoring

Suppose that a monitoring mechanism can be established at a fixed cost c'_f which is independent of the penalty imposed for cheating. It is obvious that any penalty in excess of b will be sufficient to stop all followers from cheating. Thus the fixed cost c'_f may be identified as the total transaction cost associated with monitoring:

$$t' = c'_f. \tag{3.19}$$

The transaction cost associated with manipulation has two components—just as it did in Chapter 2. The direct cost is the cost of manipulation, and the indirect cost is the loss of material co-ordination due to residual cheating. Both these costs depend on the intensity of manipulation. The minimum transaction cost is

$$t^e = v_0 - u^e = c_f + 2(abc_v)^{\frac{1}{2}}. \tag{3.20}$$

The leader chooses manipulation if

$$t^e < t'$$

i.e. if

$$c_f < c'_f - 2(abc_v)^{\frac{1}{2}}. \tag{3.21}$$

The concept of transaction cost used here resembles the familiar concept derived from institutional economics quite closely. It reflects the costs of enforcing contracts between followers. The novelty is that moral manipulation is included within the transactions technologies, as well as monitoring. The predictions of the theory are much richer than those of ordinary transaction cost theory because they include the condition for choice of technique (3.21) as well as the optimal manipulation functions (3.15) whose comparative static properties were examined, in a somewhat different context, in Chapter 2.

Monitoring costs are likely to be high wherever the default is concerned with the quality rather than the quantity of a good. When goods are easy to inspect, it is difficult to deceive the buyer about the quality, but in other cases poor quality may come to light only when the buyer attempts to utilize the good later on. Information asymmetries mean that the seller may have privileged information relating to the quality of the good, and he can use this to supply worthless 'duds' or 'lemons' instead of 'first-quality' items (Akerlof 1970).

In the context of the previous discussion, for example, the quality of specialist professional services is often difficult for the lay customer to judge. Monitoring costs are therefore high because another specialist may have to be called in to give a 'second opinion'. Moral manipulation is therefore particularly suitable in this case. Where international trade in relatively standardized products such as fabricated metals is concerned, however, monitoring costs are relatively low, while moral manipulation may be particularly difficult (for reasons of communication and cultural heterogeneity alluded to above). This is a case where monitoring is likely to be used instead.

3.6. A General Theory of One-off Encounters

Consider now the general case $k \neq 0$, in which the material gain to cheating depends upon whether the other party is expected to cheat. If $k > 0$ there is an incentive to reciprocate—in the sense that the material cost of honesty is less if the other party is honest too. In the case of teamwork (discussed in Chapter 7), for example, the cost of dedication is lower if the other party is dedicated too. More generally, as k increases there is a growing resemblance to the Assurance game described in Section 3.1. If $k < 0$, on the other hand, the incentive to be honest is greater if the other party cheats. In negotiations, for example, where an honest party compromises but a cheat does not, it may pay to compromise with a hard-liner, but to take advantage of a compromiser by taking a hard line oneself. As k decreases, there is a growing resemblance to the game of Chicken described earlier on.

In the case $k > 0$ there may be an incentive problem even if $b < 0$. If one party is certain that his partner will be honest, he will be honest too. But if $b + k > 0$, then if he thinks his partner will cheat he will cheat himself. Thus an incentive problem exists if either $b > 0$ or $b + k > 0$ (or both).

The derivation of the crime rate for the general case is illustrated geometrically in Fig 3.2, and the algebraic solutions are reported

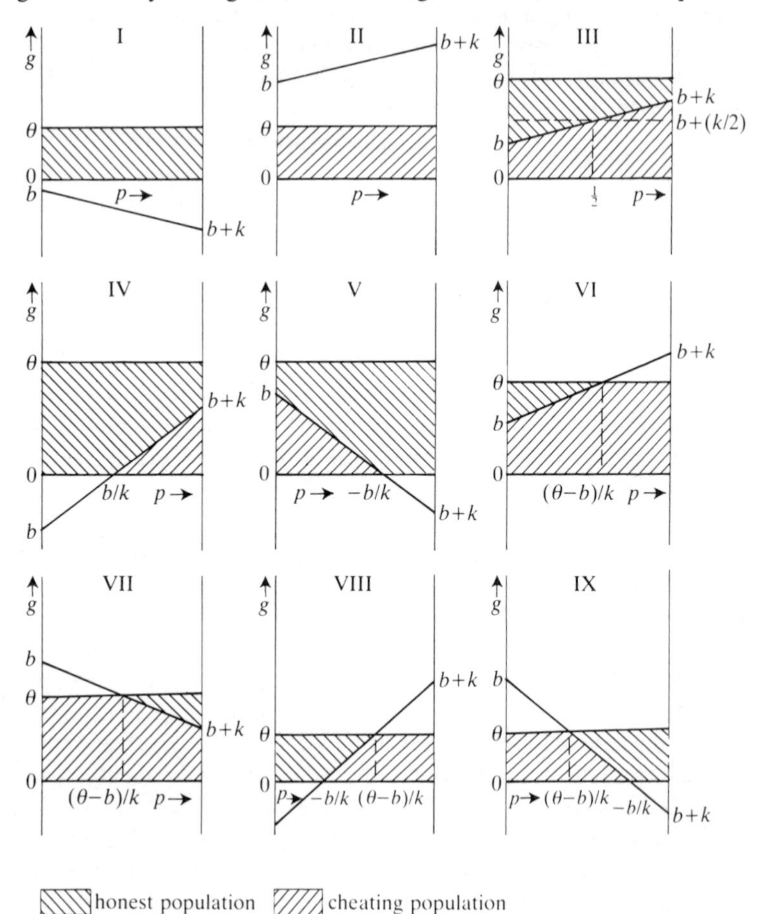

FIG. 3.2 Derivation of the crime rate for a uniform bivariate distribution of sensitivity and confidence

Note: q is measured by the ratio of the cheating population to the total population.

in Table 3.5 (column 4). The calculations relate to the uniform bivariate distribution (3.11). Since the group population is evenly dispersed in $s - p$ space, and guilt is proportional to s, with proportionality factor θ, the population is also evenly distributed over a rectangle of unit width and height θ in $g - p$ space. The linear inequality (3.1) divides this rectangle into two segments, the upper one corresponding to honest behaviour and the lower one to cheating. The crime rate is then measured by the area of the lower segment, normalized by the total area of the rectangle, θ.

The geometry of the partitioning is governed by where the inequality line intersects the vertical axis on each side of the rectangle. There are three possibilities on the left-hand axis that need to be distinguished, namely

$$b \leqslant 0, b \geqslant \theta, \text{ and } 0 < b + k < \theta.$$

Likewise there are three possibilities on the right-hand axis that must be distinguished:

$$b + k \leqslant 0, b + k \geqslant \theta, \text{ and } 0 < b + k < \theta.$$

Since the value of k is unconstrained, the left-hand possibilities are independent of the right-hand possibilities, so that altogether $3 \times 3 = 9$ cases need to be considered. Each case is identified by the roman numeral shown in the relevant quadrant of Fig. 3.2.

Of the nine regimes, only seven afford positive returns to manipulation (see Table 3.5). In regime I there is no incentive to cheat at all, and so there is no advantage to manipulation. In regime II, by contrast, the incentive to cheat is so great, and the intensity of manipulation so low, that not even the most sensitive individual will be affected by a marginal increase in manipulation and so returns to manipulation are zero in this case as well.

In regimes VI and VII, which border on II (see Fig. 3.3) returns, though positive, are constant. In regime VI increased manipulation stimulates honesty amongst less confident individuals—in other words, the heightened intensity of guilt encourages people to be honest at increasingly high levels of scepticism about the integrity of the other party. Regime VII represents the converse case—intensifying manipulation encourages people to become honest at increasingly high levels of optimism about the other party (it discourages them, in other words, from taking advantage of 'suckers'). Because returns are constant in each case, manipulation, if worth

TABLE 3.5. *Impact of the intensity of manipulation on the crime rate under a uniform bivariate distribution of sensitivity and confidence*

Type	b	Bounds					
		$b+k$	k	q	dq/θ	$d^2q/d\theta^2$	Returns
I	$b \leqslant 0$	$b+k \leqslant 0$	–	0	0	0	zero
II	$b \geqslant 0$	$b+k \geqslant 0$	–	1	0	0	zero
III	$0<b<\theta$	$0<b+k<\theta$	$-\theta \leqslant k \leqslant \theta$	$[b+(k/2]/\theta$	$-[b+(k/2]/\theta^2 \leqslant 0$	$(2b+k)/\theta^3 \geqslant 0$	diminishing
IV	$b \leqslant 0$	$0<b+k<\theta$	$k \geqslant 0$	$(b+k)^2/2k\theta$	$-(b+k)^2/2k\theta^2 \leqslant 0$	$(b+k)^2/k\theta^3 \geqslant 0$	diminishing
V	$0<b<\theta$	$b+k \leqslant 0$	$k \leqslant 0$	$-b^2/2k\theta$	$b^2/2k\theta^2 \leqslant 0$	$-b^2/k\theta^3 \geqslant 0$	diminishing
VI	$0<b<\theta$	$b+k \geqslant 0$	$k \geqslant 0$	$1-[(\theta-b)^2/2k\theta]$	$[(b^2/\theta^2)-1]/2k \leqslant 0$	$-b^2/2k\theta^3 \leqslant 0$	increasing
VII	$b \geqslant \theta$	$0<b+k<\theta$	$k \leqslant 0$	$1+[(\theta-b-k)^2/2k\theta]$	$\{1-[(b+k)^2]\}/2k \leqslant 0$	$(b+k)^2/2k\theta^3 \leqslant 0$	increasing
VIII	$b \leqslant \theta$	$b+k \geqslant \theta$	$k \geqslant 0$	$1-[(\theta/2)-b/k]$	$-1/2k$	0	constant
IX	$b \geqslant \theta$	$b+k \leqslant 0$	$k \leqslant 0$	$[(\theta/2)-b]/k$	$1/2k$	0	constant

Note: The values of q_1, $dq_1/d\theta$, and $d^2q_1/d\theta^2$, apply, where appropriate, only for $k \neq 0$. For the case $k=0$ see earlier results.

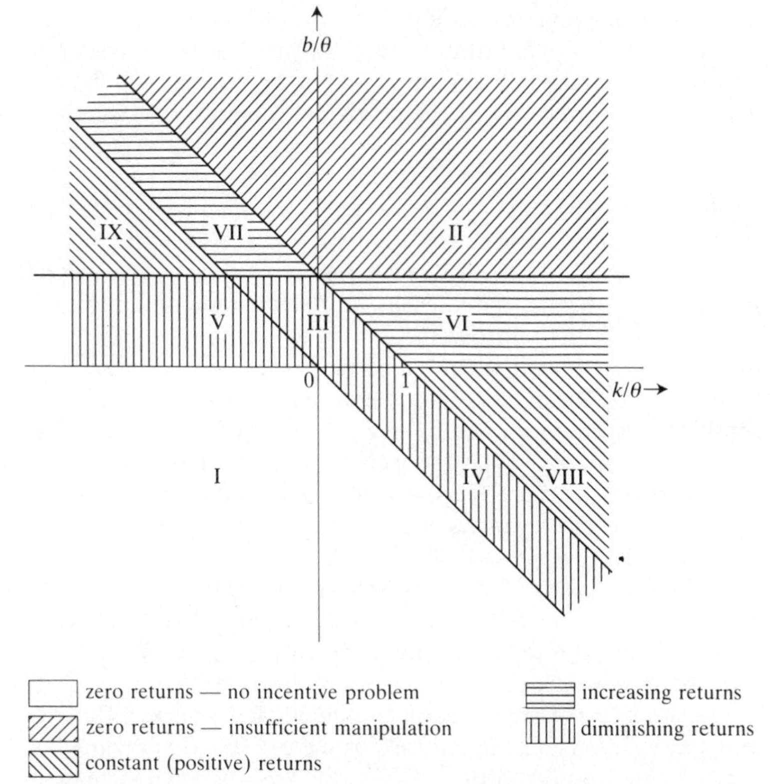

FIG. 3.3 Returns to manipulation with a bivariate uniform distribution of sensitivity and confidence

undertaking, is worth pursuing until the group as been manipulated into a different regime where diminishing returns set in.

Regimes VIII and IX are interesting and unusual because they involve increasing returns. The basic mechanism is the same as in regimes VI and VII respectively, but the sensitivity of behaviour to the perceived integrity of the other party is particularly high. An increase in manipulation moves people across the honesty margin at an accelerating rate.

Only the three remaining regimes—III, IV, and V—involve diminishing marginal returns. They all have one thing in common—the intensity of manipulation θ exceeds the maximum of b and $b + k$. This guarantees that the guilt of the most sensitive person exceeds their incentive to cheat whether they are optimistic or pessimistic about the behaviour of their partners. These three regimes are the only ones where an interior optimum may be observed. The optimizing level of manipulation, the resultant crime rate, and the net gain from maniuplation in each case are reported in Table 3.6. The comparative statics of the respective optima are explored in Table 3.7.

The results reported in these two tables are qualitatively similar to those reported earlier. The results for $k = 0$ generalize in an intuitively obvious fashion. In each case an increase in the cost of manipulation reduces the intensity of manipulation and increases the resultant crime rate. An increase in the incentive to cheat, b, increases the optimal intensity of manipulation and raises the crime rate too. The most significant new result is that the probability-sensitivity of the incentive to cheat, k, has the same qualitative effect, for given b, as does the incentive b itself. However, while the directions of the effects are the same, the magnitudes are different—the discrepancies depend on the particular parameter values in each case.

Table 3.8 indicates that the gains from manipulation are greater the lower are the fixed cost and the marginal cost of manipulation, c_f and c_v, and the lower are the incentive to cheat, b, and the probability-sensitivity of the incentive to cheat, k.

The situation considered earlier in this chapter—namely $k = 0$—may be regarded as a special case of regime III. This regime, as Fig. 3.3 indicates, lies at the 'heart' of the analysis. As θ increases, the economy converges on the borders of regime III. The rays shown in Fig. 3.4 illustrate this process. The arrow on each ray indicates the movement of the economy from its initial position towards the origin as the intensity of manipulation is increased. All points, other than those in regime I, move towards the origin through one of the regimes of diminishing returns (indicated by the vertical hatching in the figure). The pattern of transitions involved is formally presented in Table 3.9. The only ambiguity in the transition process concerns regime II. For $k > 0$, regime II switches to regime III indirectly via regime VI, whereas for $k < 0$, the transition is via regime VII. In the special case $k = 0$ the transition is direct. Note that as θ increases the economy never moves from one diminishing-returns regime to another. Only one of these regimes is relevant for any given combination of b and k (i.e. for any pattern of probability-dependent incentive to cheat).

TABLE 3.6. *Interior optima for the three regimes exhibiting diminishing returns to manipulation*

Regime	θ^e	q^e	$\Delta u = u(\theta^e) - u(0)$
III	$\{a + [b + (k/2)]/c_v\}^{\frac{1}{2}}$	$\{[b + (k/2)]c_v/a\}^{\frac{1}{2}}$	$a - c_f - 2\{ac_v[b + (k/2)]\}^{\frac{1}{2}}$
IV	$(a/2kc_v)^{\frac{1}{2}}(b + k)$	$(c_v/2ka)^{\frac{1}{2}}(b + k)$	$a - c_f - (2ac_v/k)^{\frac{1}{2}}(b + k)$
V	$(-a/2kc_v)^{\frac{1}{2}}b$	$(-c_v/2ka)^{\frac{1}{2}}b$	$a - c_f - (-2ac_v/k)^{\frac{1}{2}}b$

TABLE 3.7. *Comparative statics of the interior optima: the intensity of manipulation and the crime rate*

Impact	Regime		
	III	IV	V
$d\theta^e/dc_v$	$-\frac{1}{2}\{a[b+(k/2)]/c_v^3\}^{\frac{1}{2}} \leqslant 0$	$-\frac{1}{2}(a/2kc_v^3)^{\frac{1}{2}}(b+k) \leqslant 0$	$-(b/2)(-a/2kc_v^3) \leqslant 0$
$d\theta^e/db$	$\frac{1}{2}\{a[b+(k/2)]/c_v\}^{\frac{1}{2}} \geqslant 0$	$(a/2kc_v)^{\frac{1}{2}} \geqslant 0$	$(-a/2kc_v)^{\frac{1}{2}} \geqslant 0$
$d\theta^e/dk$	$\frac{1}{4}\{a/[b+(k/2)]/c_v\}^{\frac{1}{2}} \geqslant 0$	$\frac{1}{2}(a/2kc_v^3)^{\frac{1}{2}}(k-b) \geqslant 0$	$(b/2)(-a/2k^3c_v)^{\frac{1}{2}} \geqslant 0$
dq^e/dc_v	$\frac{1}{2}\{[b+(k/2)]/ac_v\}^{\frac{1}{2}} \geqslant 0$	$-\frac{1}{2}(b+k)/2kac_v)^{\frac{1}{2}} \geqslant 0$	$(b/2)(-2kac_v)^{\frac{1}{2}} \geqslant 0$
dq^e/db	$\frac{1}{2}\{[c_v/\mathrm{a}(b+k/2)]\}^{\frac{1}{2}} \geqslant 0$	$(c_v/2ka)^{\frac{1}{2}} \geqslant 0$	$(-c_v/ka)^{\frac{1}{2}} \geqslant 0$
dq^e/dk	$\frac{1}{4}\{[c_v/\mathrm{a}[b+(k/2)]\}^{\frac{1}{2}} \geqslant 0$	$\frac{1}{2}(c_v/2k^3a)^{\frac{1}{2}}(k-b) \geqslant 0$	$(b/2)(-c_v/2k^3a)^{\frac{1}{2}} \geqslant 0$

TABLE 3.8. *Comparative statics of the interior optima: the gains from manipulation*

Impact	Regime		
	III	IV	V
$d\Delta u/dc_f$	-1	-1	-1
$d\Delta u/dc_v$	$-\{a[b+(k/2)]/c_v\}^{\frac{1}{2}} \leqslant 0$	$-(a/2kc_v)^{\frac{1}{2}}\ (b+k) \leqslant 0$	$-b(-a/2kc_v)^{\frac{1}{2}} \leqslant 0$
$d\Delta u/db$	$-\{ac_v/[b+(k/2)]\}^{\frac{1}{2}} \leqslant 0$	$-(2ac_v/k)^{\frac{1}{2}} \leqslant 0$	$-(-2ac_v/k)^{\frac{1}{2}} \leqslant 0$
$d\Delta u/dk$	$-\{ac_v/4[b+(k/2)]\}^{\frac{1}{2}} \leqslant 0$	$-(ac_v/2k^3)^{\frac{1}{2}}\ (k-b) \leqslant 0$	$-b(-ac_v/2k)^{\frac{1}{2}} \leqslant 0$

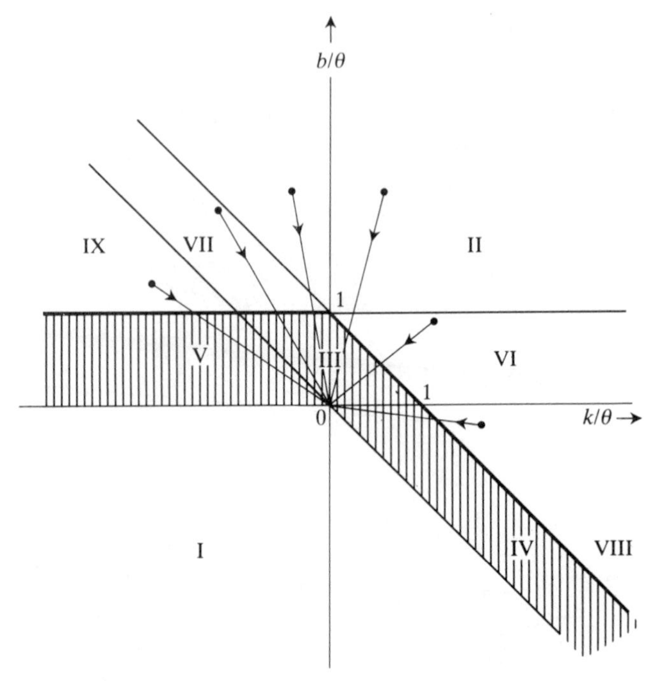

FIG. 3.4 Manipulating the transition between regimes

TABLE 3.9. *Transitions between regimes induced by an increasing intensity of manipulation*

From	To
II	(VI or VII) III
VI	III
VII	III
VIII	IV
IX	V

3.7. Announcement Effects

An important aspect of leadership is the ability to influence not only the moral attitudes of followers but also their beliefs. In the preceding analysis the leader has operated only in the moral dimension. He has modified the distribution of the population in $g - p$ space by operating only along the g axis.

In the special case $k = 0$—as exemplified by bilateral trade—the p dimension is, in fact, of no consequence, for the incentive to cheat is independent of whether other people are believed to be honest or not. When $k \neq 0$, however, manipulating beliefs can make an important contribution to economic performance. The larger the absolute value of k, the greater is the potential contribution.

Consider, therefore, a leader who can announce, before encounters commence, what he believes the crime rate will be. It is assumed that the followers trust his judgement sufficiently to take this as an accurate prediction. The same announcement is made to all members of the group, so that it has the characteristics of a public good. Since the subject of the announcement is the crime rate for the group, rather than for any specific individual, the leader is engineering the reputation of the group. It is, in fact, the reputation of the group with itself that is being affected.

When moral manipulation is coupled with an announcement, the economics of manipulation is essentially the same as it would be without an announcement but with $k = 0$. In either case the incentive to cheat is the same for all members of the population. In one case it is that their probability beliefs are all the same, and in the other because their probability beliefs, though different, do not matter.

In some cases an announcement is a complete substitute for manipulation, and in other cases only a partial one. Either way, the optimal announcement is the one that minimizes the resultant incentive to cheat. If $k > 0$, so that belief in the honesty of the other party stimulates honest behaviour, then the optimal announcement is that everyone is honest—$p = 0$. If, in addition, $b \leqslant 0$, then everyone will perceive honesty as optimal, and the announcement will be self-fulfilling without further manipulation. If $b > 0$ then some additional moral manipulation will be required, but the manipulation will be less than it would be for $p > 0$.

If, on the other hand, $k < 0$, so that belief in the honesty of the other party discourages honesty, then the optimal announcement is that everyone cheats—$p = 1$. If, in addition, $b + k \leqslant 0$, then everyone again perceives honesty as optimal—although for completely different reasons. This announcement strategy is somewhat bizarre because it works by being self-denying. It could not be used repeatedly on a group of intelligent followers. Because the present chapter is concerned with one-off encounters, however, this is not a relevant objection at this stage. It is clear, though, that in the recurrent case announcements can deal with $k > 0$ more readily than with $k < 0$. The former requires a self-validating announcement of honesty whilst the latter depends on a self-denying announcement of cheating which will face insuperable credibility problems.

The details of the optimal announcement strategies are presented in Table 3.10. It is assumed that each announcement is coupled with an optimal degree of moral manipulation. The optimal manipulation is less when an announcement has been made than when it has not. The cases reported in lines 1 and 3 require no moral manipulation at all. Those in lines 2 and 4 require additional manipulation, and the results reported assume that this manipulation is cost-effective. Indeed, it is a condition for the announcement to be economic in these cases that manipulation is economic too, because the announcement alone is insufficient to induce anyone to be honest.

The value of the announcement strategy is reported in the final column of the table. The value is higher the higher are the gains from co-ordination, the lower are the cost factors c_f and c_v (where they are relevant) and the lower is the fixed cost of the announcement, c_a. In each case the value of the announcement is greater the larger the absolute value of k. The results indicate that in the general case where the guilt requirement is probability-dependent, the leader's skill in publicizing plausible predictions (as reflected in a low cost of announcement c_a) is an important influence on group performance.

An example of the use of announcements in an Assurance game is where a team leader motivates co-operation by promoting the belief that other team members are highly motivated too. Leaders often publicly praise the commitment and loyalty of their team members, and although there are several reasons for doing so, one important reason is that members who hear this praise may be more likely to believe that colleagues whom they cannot observe are also working hard.

The use of announcements to mitigate conflict in a Chicken game is the advice given to motorists to assume that other drivers always believe that they are in the right. This discourages the belief that other drivers can be intimidated into giving way, and so encourages careful non-aggressive driving.

Because everyday life within a group consists of a mixture of

TABLE 3.10 *Announcement strategies and their value*

Reference	k	$b, b+k$	Relevant diminishing returns regime	Announced p-value	Optimal θ	q	u_a	Gains to announcement compared to pure moral manipulation
1	$k>0$	$b\leqslant 0$	IV	0	0	0	v_0-c_a	$(2ac_v/\mathrm{k})^{\frac{1}{2}}(b+k)+c_f-c_a$
2	$k>0$	$b\leqslant 0$	III	0	$(ab/c_v)^{\frac{1}{2}}$	$(bc_v/a)^{\frac{1}{2}}$	$v_0-c_a-c_f-2(abc_v)^{\frac{1}{2}}$	$2(ac_v)^{\frac{1}{2}}([\mathrm{b}+(k/2)]^{\frac{1}{2}}-b^{\frac{1}{2}})-c_a$
3	$k<0$	$b+k\leqslant 0$	V	1	0	0	v_0-c_a	$(-2ac_v/k)^{\frac{1}{2}}b+c_f-c_a$
4	$k<0$	$b+k>0$	III	1	$[a(b+k)/c_v]^{\frac{1}{2}}$	$[(b+k)c_v/a]^{\frac{1}{2}}$	$v_0-c_a-c_f$ $-2[a(b+k)c_v]^{\frac{1}{2}}$	$2(ac_v)^{\frac{1}{2}}\{[b+(k/2)]^{\frac{1}{2}}$ $-(b+k)^{\frac{1}{2}}\}-c_a$

different types of game, a leader who wishes to spread a consistent view of other people faces a delicate trade-off. He must build up confidence in certain attributes but undermine confidence in others Most political leaders seem to opt for confidence-building across the board, possibly because they themselves like the world of optimistic delusion thereby created. A more realistic approach is to emphasize that people may well be trustworthy in fairly small matters, particularly where they face familiar situations in which the moral dilemmas are easily recognized, but that where large matters are concerned it is necessary to be prudent and to select a partner with care. In the present model individuals cannot choose their partners, and so this aspect of business culture does not actually arise, but in Chapter 8, where reputable intermediators appear, it is shown that the leader can partially reconcile conflicting aims by focusing trust selectively on a business élite.

3.8. Summary

This chapter has extended the model of manipulation expounded in Chapter 2 to pairwise encounters within a group. Many of the results obtained in Chapter 2 generalize naturally—up to a point. That point is marked by the PD game, where each player's benefit from cheating is independent of how his partner behaves. This game is exemplified by bilateral trade. The transformation of a PD into a game of Harmony is effected entirely by making individuals more trustworthy. It is not necessary to persuade them that other people are trustworthy too.

In all other cases the material incentive to cheat depends upon how the partner behaves. In games of Assurance, co-ordination is effected both by encouraging integrity and building up confidence in others. In games of Chicken, co-ordination involves making people honest and encouraging them to distrust other people.

Encouraging integrity in trade is easiest when the seller's valuation of the product is considerably below the buyer's, for it is the seller's valuation that determines the material incentive to cheat. This condition is most likely to be satisfied for customized products. It is least likely to be satisfied for standardized and highly 'liquid' products such as land.

The alternative to moral manipulation of traders is monitoring backed up by the legal system. Enforcing contracts by law is most effective when uncertainty about the quality of the product is low and when manipulation costs are high. High manipulation costs can be due to uninspiring leadership, high media costs, and inappropriate traditions (as emphasized in Chapter 2), or to a relatively high level of seller valuation, as indicated above. Reasoning of this kind makes it possible to derive a wide range of predictions about the culture and institutions governing trade. But to achieve greater realism in the predictions it is necessary to relax some of the underlying assumptions, and this is the agenda for subsequent chapters.

4

The Causes of Catastrophe

4.1. Introduction

This chapter and the next extend the analysis of Chapter 3 in two distinct, though complementary ways. This chapter looks at repeated encounters, and contrasts them with the one-off encounters studied in Chaper 3. The context is, however, slightly different from that usually studied in the theory of repeated games (Friedman 1986). In the present case, people trade repeatedly within the same group rather than with the same specific partner. People can learn from their trading experiences, as they do in conventional theory, although the learning is assumed to take a rather simple unsophisticated form. This simple learning process is in line with the earlier assumption that people work with an oversimplified model of their environment anyway (see Section 3.2). Learning creates reputation, but in the present context reputation adheres not to individuals but to the group as a whole.

Once again trade stands out as a special case. Because the incentive to default is independent of whether others default, information about whether others have previously defaulted is irrelevant to the trader. A trader facing a recurrent PD will not normally revise his behaviour in the light of experience. This very strong result arises, however, only because participation in trade is assumed to be compulsory. When participation is made voluntary in Chapter 5 this result no longer applies, since if dishonesty is believed to be rife, members of a group will avoid participating in trade altogether. Nevertheless, the result highlights an important point: the structure of material incentives faced by players determines whether, in repeated encounters, learning is worthwhile or not.

Where learning is worthwhile, it is important to distinguish between the case where optimism about other people's honesty encourages honest behaviour, and the case where the reverse applies. The first case is far more common than the second, and this is fortunate, because it affords a virtuous circle—or positive feedback loop—which the other does not. If the leader can just 'prime the pump' by getting a few more people to be honest, then learning effects will increase optimism and cause a further improvement in integrity. Complete honesty in encounters may be achieved by only minimal effort from the leader. The leader simply needs to know the critical level of manipulation required to prime the pump in the first case.

The second case, by contrast, leads to the bizarre result that as the leader stimulates honesty so the feedback of information induces more default. This second case is likely to lead to an alternating cycle of honesty and cheating, which is broken only if people realize the distinctive pattern that has emerged and alter their own behaviour because they anticipate it.

The logic of repeated encounters is relatively complex, and so the analysis is developed in a number of stages. Section 4.2 sets out the basic assumptions about learning. Section 4.3 looks at the dynamics of group reputation. It shows how, with a given distribution of moral sensitivity, the current crime rate is determined recursively from last period's crime rate. By equating the current crime rate to the previous crime rate, the steady-state equilibrium crime rate can be derived. This equilibrium crime rate is governed, as before, by the interplay of the intensity of manipulation and the material incentive to cheat. There is, in fact, a potential multiplicity of equilibria, but discussion of this problem is deferred until later in the chapter. To begin with, parameter values (such as the size of the material incentive to cheat) are restricted in order to guarantee a unique interior equilibrium.

Section 4.4 deduces the optimal leadership strategy under these conditions. The method is exactly analogous to that of Chapter 3. The only difference is in the relation between the equilibrium crime rate and the intensity of manipulation. This difference encourages a different leadership strategy—except in the case of trade, as indicated above. Section 4.5 compares the optimal leadership strategy with the monitoring alternative.

Multiple equilibria are considered in Section 4.6, and stability is discussed in Section 4.7. The analysis of stability raises the alarming possibility that under certain conditions the economy will degenerate into comprehensive mutual cheating once a critical level of integrity fails to be met. Once the crime rate falls below a critical level governed by the material incentive to cheat, reputation effects plunge the economy towards an equilibrium of mutual cheating at an accelerating rate.

On the more positive side, however, it is also possible that under other conditions the economy will converge on comprehensive mutual honesty provided that a critical level of the crime rate is not exceeded. This is the basis of the virtuous circle referred to earlier.

4.2. Learning from Experience

An important theme in modern economics is that intelligent people will learn from their mistakes. Information on the outcomes of previous encounters will be fed back to modify expectations about future encounters. Their modified expectations will encourage adaptive behaviour. The question then arises as to whether this process of adaption will settle down over time to generate an equilibrium. If so, will this equilibrium be stable—or will the system diverge from it permanently if subjected to a small environmental change?

A common prejudice amongst economists is that greater provision of information accelerates adjustment towards a stable equilibrium. It is assumed that with greater information the economy approximates more closely to the ideal of a competitive market economy. This argument implicitly assumes a dependable framework of law. Once this assumption is relaxed, the intuition that underlies the argument has only very limited validity. Under certain conditions information feedback can easily transform a desirable equilibrium into a social catastrophe.

The preceding chapter analysed a one-period economy in which each individual had a single encounter with another. The present chapter allows the economy to operate for an unlimited number of periods. In each period information is available on the previous period's outcome. As in the previous analysis, this information relates to the reputation of the group with itself. Each individual has access to a single statistic—the crime rate averaged across the group as a whole.

Because of the follower's need to simplify decision problems, it is assumed that, apart from a knowledge of his own rewards, the crime rate is the only information he has available. He cannot recall the specific results of previous encounters with particular individuals. This is a plausible assumption for a large group, in which the expected frequency of meeting any particular individual a second time is small.

It is also assumed that the latest information completely supersedes earlier results. Thus group reputation is based on just the latest crime rate, and not, for example, a weighted average of previous rates. Although this learning pattern is non-Bayesian and relatively unsophisticated, it reduces the memory power and computing power that the individual requires.

Finally, it is assumed that the intensity of manipulation remains constant over the lifetime of the group. The leader cannot fine-tune manipulation by adjusting on a period-by-period basis to compensate for followers' changing beliefs. All of the leader's strategy formulation is telescoped into the first period, in which he decides whether to make an announcement and what intensity of manipulation to apply.

The present analysis does not entirely supersede the analysis of one-off encounters in Chapter 3. In a rapidly changing environment, information about the past may be totally irrelevant to predicting the future. If each period brings its own unprecedented situations, then from an informational point of view each period stands alone, with previous experience being irrelevant and its own implications for the future being negligible. The previous analysis may still be applied to analyse behaviour over time, therefore, but only when the passage of time brings a sequence of disconnected novel situations.

The previous analysis is also relevant in establishing initial conditions for the present analysis. In the first of the recurrent rounds of encounters, followers have no relevant experience to draw upon. They must therefore rely on subjective probabilities to get their sequence of encounters going. The previous analysis suggests two main possibilities—first, that subjective probabilities are uniformly distributed across the population, and secondly that they are homogenized through an announcement by the leader—and both of these are considered here.

4.3. Dynamics of the Crime Rate

Let $F_1(s,p)$ be the initial bivariate distribution of sensitivity and probability across the group. It is assumed that with no announcement the distribution is uniform, as before:

$$F_1 = sp \qquad 0 \leqslant s, p \leqslant 1, \tag{4.1}$$

whilst with an announcement it is

$$F_1^a = \begin{cases} 0 & p < \varphi \\ s & p \geqslant \varphi, 0 \leqslant s \leqslant 1 \end{cases} \tag{4.2}$$

where φ is the announced probability. The initial beliefs generate an observed crime rate which can be derived from the analysis of Chapter 3. This crime rate forms the basis for beliefs in the next period, and so on. In subsequent periods the distribution is

$$F_t = \begin{cases} 0 & p < q_{t-1} \\ s & p \geqslant q_{t-1}, 0 \leqslant s \leqslant 1 \end{cases} \tag{4.3}$$

(t = 2, 3, ...) where it is readily deduced that the corresponding crime rate is

$$q_t = \begin{cases} 0 & q'_t < 0 \\ q'_t & 0 \leqslant q'_t \leqslant 1 \\ 1 & q'_t > 1 \end{cases} \tag{4.4}$$

where q'_t follows the first order linear recurrence relation

$$q'_t = (b + kq'_{t-1})/\theta. \tag{4.5}$$

The derivation of the crime rate is illustrated in Fig. 4.1 for the special case $b>0$, $k \geqslant 0$. In the left-hand quadrant the height of the schedule BB′ indicates the critical level of guilt required to induce honesty. The height of the horizontal schedule AA′ indicates the

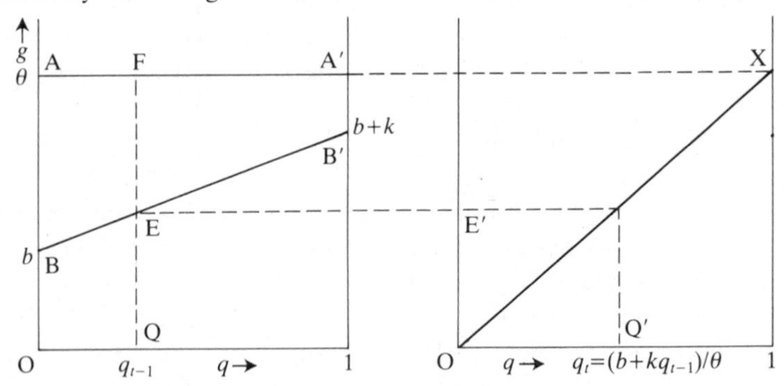

FIG 4.1 Determination of crime rate when $b > 0$, $k \geqslant 0$

intensity of manipulation. The crime rate of the previous period is measured by the horizontal intercept OQ. The current period's crime rate is measured by the ratio of the segment EQ—representing those who cheat—to the segment FQ—representing the entire population, evenly distributed by sensitivity across the segment. This ratio is projected on to the horizontal axis in the right-hand quadrant using the ray OX, whose slope is given by the intensity of manipulation, θ. The projection indicates that the crime rate in the current period is measured by the intercept QQ′.

An equilibrium, if it exists, must satisfy the condition.

$$q_t = q_{t-1} \equiv q^* \qquad 0 \leqslant q^* \leqslant 1. \tag{4.6}$$

In the special case $b > 0$, $k \geqslant 0$, combining (4.4)–(4.6) gives

$$q^* = \begin{cases} b/(\theta - k) & \theta \geqslant b + k \\ 1 & \theta < b + k. \end{cases} \tag{4.7}$$

The equilibrium is illustrated in Fig. 4.2. The upper quadrant (A) represents the case $\theta \geqslant b + k$, which corresponds to regime III of Chapter 3. The schedule QQ′, derived from (4.4) and (4.5), indicates the dependence of the current crime rate on the previous one. The equilibrium crime rate q^* is determined by the intersection E of QQ′ with the diagonal OZ.

The stability of the equilibrium can be seen by tracing out the trajectory of the crime rate, beginning with q_1. Suppose, for example, that the leader initially announces $\varphi = 0$, giving $q_1 = b/\theta$. When values of the crime rate are read off from the vertical axis then this first outcome is represented by the point Q. Moving to the next period, Q is translated to the diagonal to produce a reading for the last period's crime rate along the horizontal axis. Given that during the last period followers were expecting no crime, but observed that some crime occurred, some of the less sensitive individuals will begin to cheat. Reading along up to the point Q_2 shows that, as a result, the new crime rate is higher than the old. In the next period the further modification of expectations generates, in turn, a crime rate indicated by Q_3. As the process continues, the system trajectory converges on E.

When $k > 0$ the announcement $\varphi = 0$ is, in fact, the rational one-period strategy. But suppose that instead the leader had announced $\varphi = 1$. This would generate the higher first-period crime rate $q_1 = (b + k)/\theta$, indicated by the point Q′. Since this forecast turns out to be too pessimistic, some cheats subsequently

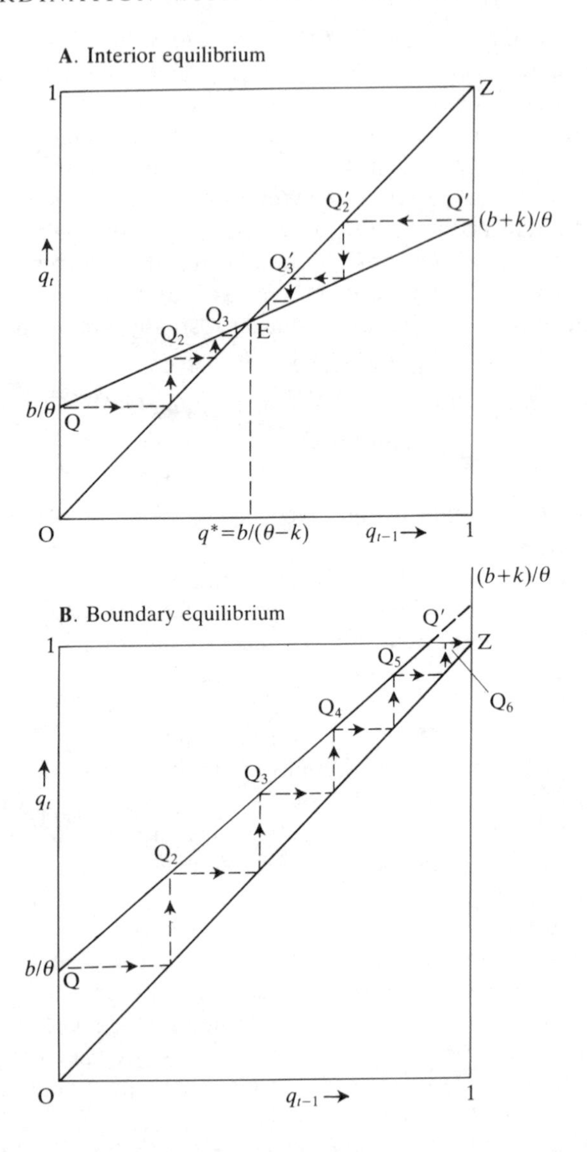

FIG. 4.2 Equilibrium in recurrent encounters

become honest, and the crime rate reduces steadily, through Q_2', Q_3', and so on, to the equilibrium at E. Thus the equilibrium is stable from whatever direction the adjustment process is begun.

The lower quadrant (B) illustrates the case $\theta < b + k$, in which there is an equilibrium at $q^* = 1$. The schedule QQ′ Z represents the dynamics of the crime rate. In the case illustrated, where $\theta > b$, the segment QQ′ slopes upwards from left to right. If $\theta < b$ then the segment QQ′ coincides with the top of the unit box, but the key result is the same—the intersection with the diagonal OZ occurs only at Z, which is therefore the unique equilibrium point.

The stability of Z is illustrated by the convergence of the system trajectory beginning at Q. This trajectory traces out the time path of the crime rate following an initial announcement $\varphi = 0$ which turns out to be totally false. The announcement is false because the intensity of manipulation θ is so small that once people realize that some are cheating, more wish to cheat, until the expectation that everyone will cheat becomes self-fulfilling.

4.4. Optimal Manipulation with Recurrent Encounters

When a leader is faced with recurrent encounters his optimizing strategy is quite different from that in the one-off case. It is assumed below that the leader's preferences relate to the equilibrium situation—he is not concerned with the trajectory, and so is indifferent as to how long the equilibrium takes to reach. The costs of manipulation are now treated as recurrent costs, and it is assumed for simplicity that they assume the same values as in the one-off case.

An altruistic materialistic leader maximizes

$$u = v - c \tag{4.8}$$

where

$$v = v_0 - aq^* \tag{4.9}$$

and

$$c = \begin{cases} 0 & \theta = 0 \\ c_f + c_v\theta & \theta > 0. \end{cases} \tag{4.10}$$

Equations (4.8)–(4.10) are the recurrent analogues of equations (3.7), (3.8), and (3.10).

Continuing with the case $b > 0$, $k \geqslant 0$, it is clear that if $\theta < b + k$ then there are zero returns to manipulation, since the only equilibrium is one in which everybody cheats. Any attempt to manipulate favourable initial conditions is fruitless, for the only effect can be to delay, but not prevent, convergence on the cheating equilibrium. The search for an interior optimum can therefore be confined to $\theta \geqslant b + k$. Substituting (4.7), (4.9), and (4.10) into (4.8) and applying this condition gives

$$u = v_0 - c_f - (ab/(\theta - k)) - c_v\theta. \tag{4.11}$$

The first-order condition for a maximum is:

$$du/d\theta = ab/(\theta - k)^2 - c_v = 0, \tag{4.12}$$

whence

$$\theta^e = k + (ab/c_v)^{\frac{1}{2}} \tag{4.13.1}$$
$$q^e = (bc_v/a)^{\frac{1}{2}} \tag{4.13.2}$$
$$v^e = v_0 - (abc_v)^{\frac{1}{2}} \tag{4.13.3}$$
$$u^e = v_0 - 2(abc_v)^{\frac{1}{2}} - c_vk - c_f. \tag{4.13.4}$$

The second-order condition for a maximum is

$$d^2u/d\theta^2 = - 2ab/(\theta - k)^3 < 0, \tag{4.14}$$

which is always satisfied given the inequalities specified earlier.

Setting $k = 0$ in (4.13) gives the results for recurrent trade. It is striking that these results are identical with those derived for one-off trade in Chapter 3. The reason is that in the special case $k = 0$ traders do not benefit from learning since their optimal strategy is independent of the probability of cheating. Since no learning occurs, the initial outcome replicates itself indefinitely. This result does not apply in any other case.

4.5. Monitoring

Once again a comparison with monitoring can be made. The recurrence of encounters has a limited effect on monitoring strategy. The taking of hostages as penalties in a one-off encounter entirely eliminates cheating provided the penalty is of sufficient size, and the same principle applies in recurrent encounters. A penalty exceeding $b + k$ suffices to eliminate cheating whatever people believe about the integrity of others, and is therefore effective in perpetuity. Since by assumption monitoring involves a fixed cost $c_f' > 0$, but no variable cost, there is no problem in imposing a penalty of sufficient size.

The recurrent transaction cost associated with manipulation is:

$$t^e = c_f + c_vk + 2(abc_v)^{\frac{1}{2}} \tag{4.15}$$

and that associated with monitoring is:

$$t' = c'_f. \tag{4.16}$$

Manipulation is preferred to monitoring if $t^e < t'$, i.e. if:

$$c_f < c'_f - c_v k - 2(abc_v)^{\frac{1}{2}}. \tag{4.17}$$

Comparison with (3.21) shows that when $k = 0$ the choice between monitoring and manipulation is based on exactly the same criterion as before.

4.6. General Equilibrium Analysis

An important feature of recurrent trading is the possibility of multiple equilibria, and the potential for catastrophic switches between them. Equations (4.4) and (4.5) provide the key to general equilibrium analysis of the system. Three main types of equilibrium may be distinguished.

1. *An equilibrium of mutual honesty, in which the crime rate is zero.* A necessary and sufficient condition for such an equilibrium to be sustained by a finite intensity of manipulation is that the incentive to cheat is zero or negative, $b \leqslant 0$. To see this, note that $q_t = q_{t-1} = 0$ if and only if, from (4.4)

$$q'_t - q_{t-1} \leqslant 0 \tag{4.18}$$

when $q_{t-1} = 0$. Since, from (4.5),

$$q'_t - q_{t-1} = (b + (k - \theta)\, q_{t-1})/\theta, \tag{4.19}$$

we have, for $\theta > 0$ and $q_{t-1} = 0$, $b \leqslant 0$. Note that if $b > 0$ it takes an infinite intensity of manipulation to achieve this equilibrium.

2. *An equilibrium of mutual cheating in which the crime rate is unity.* From (4.4) it may be deduced that such an equilibrium requires

$$q'_t - q_{t-1} \geqslant 0 \tag{4.20}$$

when $q_{t-1} = 1$. Using (4.19) and taking $\theta > 0$ indicates that (4.20) is satisfied if and only if $\theta \leqslant b + k$. This confirms that an equilibrium of mutual cheating arises principally because the intensity of manipulation is too low.

3. *An interior equilibrium* with crime rate $q^*(0 < q^* < 1)$. From (4.4) the necessary and sufficient condition is that:

$$q'_t - q_{t-1} = 0 \tag{4.21}$$

when $q_{t-1} = q^*$. Applying (4.19) with $\theta > 0$ and $q_{t-1} = q^*$ gives:

$$q^* = b/(\theta - k). \tag{4.22}$$

The condition $q^* > 0$ implies that:

$$\text{either} \quad b > 0 \text{ and } \theta > k \tag{4.23.1}$$
$$\text{or} \quad b < 0 \text{ and } \theta < k. \tag{4.23.2}$$

The condition $q^* < 1$ implies additionally that:

$$\text{either} \quad \theta - k > b > 0 \tag{4.24.1}$$
$$\text{or} \quad \theta - k < b < 0. \tag{4.24.2}$$

Since the latter conditions imply the former corresponding condition, but not vice versa the latter are sufficient. They are also necessary in their own right. It follows that (4.24) is a necessary and sufficient condition for an interior equilibrium.

Note that (4.24.2) implies $b \leqslant 0$ and $\theta \leqslant b + k$, so that both zero and unity are equilibrium values too. This means that (4.24.2) generates a triple equilibrium. On the other hand (4.24.1) implies $b > 0$ and $b + k < \theta$, so that the q^* equilibrium is unique. Finally, note that $b = 0$, $\theta = k$ generates an infinity of equilibria along the unit interval $0 \leqslant q^* \leqslant 1$.

4.7. General Stability Analysis

One of the most remarkable features of recurrent encounters is the potential instability of the outcome. This instability stems from the probability-dependence of the incentive to cheat. When the incentive to cheat depends critically on expectations about cheating, new information about cheating in the previous period can cause a major change in behaviour. System trajectories can diverge at critical points—a crime rate slightly below the critical level can trigger a virtuous circle of reducing crime, whilst a crime rate slightly above this level triggers a vicious circle which can end with everyone cheating on everyone else.

This section examines the determinants of stability in a systematic way. Five main regimes are distinguished in terms of their dynamic properties. A feature of the least stable regime is a low intensity of manipulation. This highlights the fact that in recurrent trading, intensive manipulation not only tends to improve the equilibrium outcome, but to enhance the stability of the system too.

Consider, to begin with, the stability of the mutual honesty equilibrium. If this equilibrium is displaced by a small increase in reported cheating then it will return to mutual honesty if $b < 0$. If $b = 0$ it will only return to equilibrium if $k < \theta$. If $k = \theta$ it will remain in its disturbed state, whilst if $k > \theta$ it will diverge towards a mutual cheating equilibrium.

If the mutual cheating equilibrium is displaced by a small reduction in reported cheating then it will return to mutual cheating if $\theta < b + k$. *If* $\theta = b + k$ then it will only return to equilibrium if $k < \theta$. If $k = \theta$ then, as in the previous case, it will remain in the disturbed state, whilst if $k > \theta$ it will diverge in the direction of a mutual honesty equilibrium.

Finally, the interior equilibrium is stable whenever $k < \theta$. For $k = \theta$ any disturbance is perpetuated, whilst for $k > \theta$ an increase in reported cheating causes divergence towards mutual cheating; conversely a reduction in reported cheating causes divergence towards mutual honesty.

These results highlight the importance of the conditions $b < 0$, $b + k < \theta$, and, above all, the relation between k and θ. Whenever there is a potential instability problem it is the relation between k and θ which is crucial.

Combining the analysis of equilibrium and stability identifies the five regimes A–E listed in Table 4.1. The relations between these regimes are illustrated in Fig. 4.3. The crosses in the figure identify typical situations in each of the five regimes, whose stability properties are illustrated in detail in Fig. 4.4. Each segment of this figure represents the response function by a bold line. The response function indicates how today's crime rate depends on yesterday's crime rate; it is derived from equations (4.4) and (4.5), and the appropriate inequalities on b, k, $b+k$, and θ.

The dynamics of adjustment following an arbitrary announcement of a crime rate by the leader are illustrated by the trajectories, which are analogous to those of Fig. 4.2. Segment A of the earlier figure corresponds, in fact, to a special case of regime C, whilst segment B corresponds to a special case of regime B.

Regime A has a unique stable equilibrium of mutual honesty to which the system converges whatever the initial state. Regime B is, in a sense, the reverse of this, having a unique stable equilibrium of mutual cheating instead. Regime C has a unique and stable interior equilibrium, whilst D exhibits explosive oscillations around an unstable interior equilibrium. Finally, E exhibits a

TABLE 4.1. *Dynamics of the crime rate with repeated operation*

Regime	b	k	$b+k$	Stable equilibria	Unstable equilibria	Pattern
A	$b \leqslant 0$	—	$b+k \leqslant \theta$	0	—	Convergence to honesty from any initial state
B	$b \geqslant 0$	—	$b+k \geqslant \theta$	1	—	Convergence to cheating from any initial state
C	$b \geqslant 0$	$-\theta \leqslant k \leqslant \theta$	$b+k \leqslant \theta$	$q^* = b/(\theta - k)$	—	Convergence to q^* from any initial state
D	$b \geqslant 0$	$k \leqslant -\theta$	$b+k \leqslant \theta$	—	$q^* = b/(\theta - k)$	Explosive oscillations from any initial state except q^*
E	$b \leqslant 0$	$k \geqslant \theta$	$b+k \geqslant \theta$	0, 1	$q^* = b/(\theta - k)$	Bifurcation about an unstable equilibrium q^*

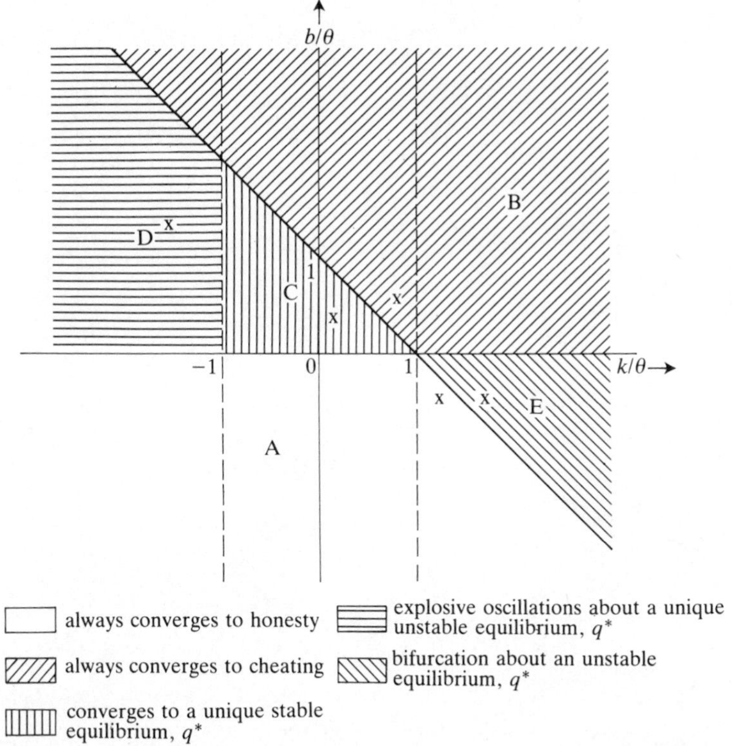

FIG. 4.3 Stability analysis for repeated operation with a uniform distribution of sensitivity

bifurcation in which the system diverges to either mutual honesty or mutual cheating, depending on which side of the unstable equilibrium crime rate the initial condition is placed.

Since the main cause of both an undesirable stable equilibrium, and of instability, is lack of manipulation, intensification of manipulation can be applied to manoeuvre the system into a more desirable regime. The consequences of intensified manipulation for various initial regimes is illustrated in Fig. 4.5.

With recurrent encounters the impact of the intensity of manipulation on the equilibrium crime rate is different from that in the one-off case. Differentiating (4.22) for the equilibrium crime rate shows that throughout regime C (where (4.22) applies) there are diminishing marginal returns to manipulation. As a result, there will normally be pressure to increase θ until regime C is entered. Then θ is optimized

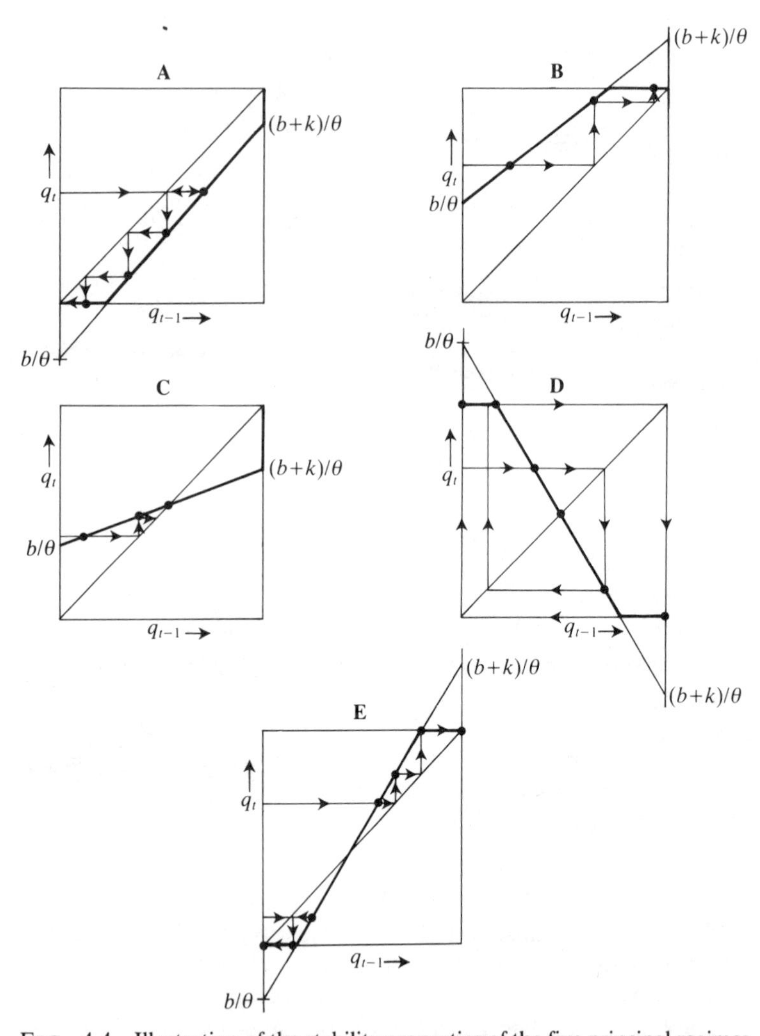

FIG. 4.4 Illustration of the stability properties of the five principal regimes

by selecting a suitable point along the relevant ray—with more intensive manipulation moving the system closer to the origin.

There are three main exceptions to this strategy. The first is a trivial one. It does not apply to regime A, where no incentive problem exists in the first place, and where returns to manipulation are zero.

Secondly, it may not apply to regime D if the leader is unconcerned

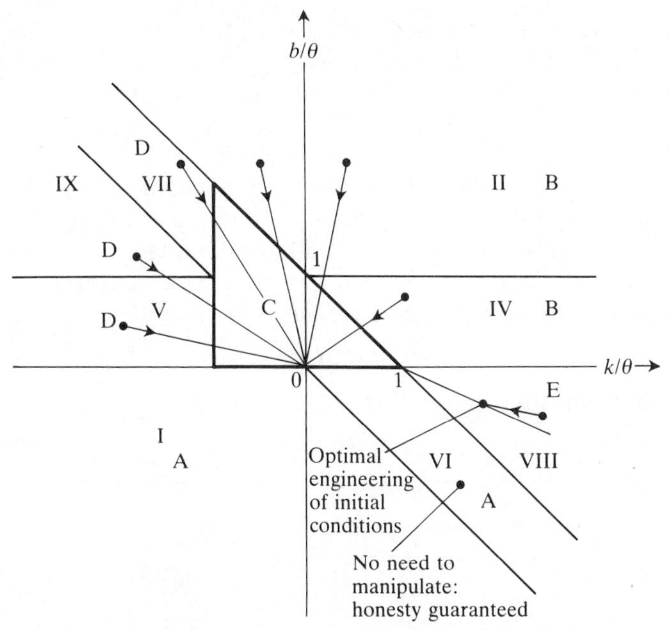

FIG. 4.5 Manipulating the system to a desirable stable state with repeated operation

Note: The area C exhibits diminishing returns to manipulation: $dq^*/d\theta < 0$, $d^2q^*/d\theta_2 > 0$.

about the oscillations in behaviour. In regime D the system alternates between total honesty and total cheating so that a two-period moving average of the crime rate is stable at one-half. This may be a better result than what can be achieved at acceptable cost by intensifying manipulation to suppress the oscillations through moving the system to regime C. In practice, of course, the persistence of such oscillations is dubious since transactors will learn from the pattern of behaviour. Nevertheless, if the consequence of learning is simply to modify the pattern of oscillation rather than suppress it, then the properties of a suitably extended moving average may remain unchanged.

The third case is the most interesting. It concerns regime E which, alone among the regimes, exhibits a limiting pattern of behaviour which is dependent on the initial conditions. In all the other regimes, the final equilibrium achieved, or the pattern of oscillation generated, is independent of the initial conditions. In regime E, by contrast, the leader who engineers an initial crime rate just below the value of the unstable interior equilibrium can rely on the system converging to honest trading. The leader can engineer this either by an announcement or by setting an appropriate intensity of manipulation. In either case the strategy will be more efficient than increasing the intensity of manipulation to the point where regime A is attained.

The manipulation strategy turns out to be particularly simple. Regime E corresponds to regime VIII of Chapter 3 so far as the one-off engineering of the initial crime rate is concerned. The relevant equation for the initial crime rate is therefore

$$q_1 = 1 - \{[(\theta/2) - b]/k\} \tag{4.25}$$

and from equation (4.22) the restriction $q_1 < q^*$ implies

$$q_1 < b/(\theta - k). \tag{4.26}$$

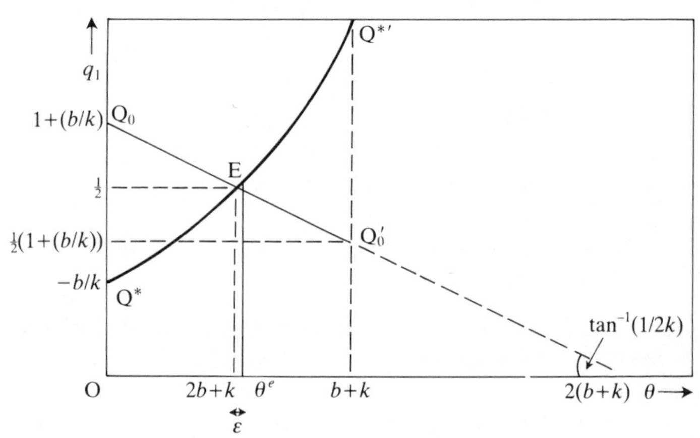

FIG. 4.6 Engineering an unstable interior equilibrium and corresponding initial crime rate, so that the equilibrium in recurrent encounters is one of mutual honesty

The solution, illustrated diagrammatically in Fig. 4.6, is

$$\theta^e = 2b + k + \varepsilon \tag{4.27.1}$$

$$q^* = \tfrac{1}{2}, \tag{4.27.2}$$

where the small displacement $\varepsilon > 0$ ensures that the strict inequality in (4.26) is satisfied. The schedule $Q_0Q'_0$ represents (4.25), which is valid over the region $\theta \leqslant b + k$, after which a different regime applies. The schedule $Q^*Q^{*\prime}$ represents the upper left-hand boundary of the region in $q_1-\theta$ space defined by (4.26).

The intersection E of Q_0Q_0' and $Q^*Q^{*\prime}$ represents the optimal value of the unstable interior equilibrium, corresponding to a value $\theta = 2b + k$, and crime rate $q^* = \frac{1}{2}$. To reduce the initial crime rate marginally below the interior equilibrium, it is sufficient to increase θ by a small amount ε. Since $Q^*Q^{*\prime}$ is upward sloping, while Q_0Q_0' is downward sloping, this raises the interior equilibrium slightly and at the same time reduces the initial crime rate. Solutions of this type for various combinations of b and k, map out the line that bisects the region E, as shown in Fig. 4.5.

4.8. Summary

With good information flow within a relatively stable environment it is appropriate to analyse behaviour in the context of a sequence of encounters. This chapter has shown that the behaviour revealed with repeated encounters is qualitatively different from that in a one-off encounter. Learning effects have a capacity to destabilize the economy. When manipulation is of only modest intensity, learning about other people's dishonesty drags the economy into a mutual cheating equilibrium. In seeking to raise the intensity of manipulation, the leader can sometimes exploit the instability of an interior equilibrium to set a modest 'pump-priming' level of manipulation which, through a virtuous circle will lead the economy to a stable equilibrium of mutual honesty. The same effect can be produced if the leader primes the pump with an optimistic prediction of the crime rate. Leaders can therefore exploit instability to their own advantage, but only if they fine-tune their strategies correctly. A small deviation from an optimal strategy can have catastrophic consequences.

These results have important implications, not only for economic performance but also for the collapse of social order and the extinction of politico-economic systems. To apply the analysis to such issues, however, it is desirable to enrich it further by considering how catastrophic effects can be amplified by changing levels of active participation in group activity. This is the subject of the next chapter.

5

Promoting Participation

5.1. Voluntary Participation

The message of the three preceding chapters has been that poor economic performance stems from a high incidence of cheating. Casual empiricism suggests, however, that while a link between cheating and performance undoubtedly exists, problems of cheating often impact on performance by a rather different route. People who expect to be cheated do not participate in group activity at all. Poor economic performance is associated not so much with actual cheating as with fear of being cheated. It is reflected in a low level of activity across the board. People who fear being attacked stay at home, so that actual crimes committed may well seriously understate the magnitude of the problem. People who cannot trust other people's workmanship engage in do-it-yourself production instead. This restricts the gains from specialization that can be achieved by a division of labour in the economy. Potential customers avoid new products because they suspect there may be a hidden flaw. Potential workers refuse to join teams because they fear that their collegues may not co-operate.

When participation is voluntary, encounters generate a two-stage game. The first stage involves a decision whether to participate, and the second whether, if participating, to be honest or not. So far the discussion has focused exclusively on the second stage, on the assumption that the decision at the first stage is always to participate. This chapter incorporates the first stage into the analysis too.

With voluntary participation, the crime rate and the participation rate are simultaneously determined (Akerlof 1970). Both depend on the material incentives and the cost of manipulation. This common dependence determines a systematic relation between them. Since both the crime rate and the participation rate are, in principle, readily measurable, the predicted relation between them can be tested in a straightforward way.

The most important implications of voluntary association only emerge when encounters are repeated. This is because when moral sensitivities differ, participation becomes a selective process. Reputation effects stimulate this selective process and, if not controlled by the leader, can precipitate catastrophe. Voluntary participation thus exaggerates the dangers highlighted in the previous chapter.

Selectivity works in the following way. Highly sensitive individuals will normally wish to participate only if they can be honest, because otherwise participation will generate intense feelings of guilt. But if they are honest they are very vulnerable to being taken for a sucker, and so their decision to participate honestly is highly sensitive to their expectation of cheating.

Less sensitive individuals, on the other hand, are more willing to participate simply in order to take advantage of others by cheating. As expectations of honesty decline, therefore, the disincentive to participate acts less on those of low sensitivity than on those whose sensitivity is high. As a result, the incidence of cheating tends to rise as expectations of cheating increase because disproportionate numbers of potentially honest participants withdraw. This leads to a self-validating vicious circle in which eventually only the cheats participate. Since cheats cannot prosper from encounters when the only possible partners are other cheats, social activity disintegrates altogether. Everyone avoids encounters and so no new information becomes available which could restore confidence. The economy sinks irreversibly into an equilibrium of complete inactivity.

Voluntary participation clearly necessitates greater sophistication in leadership strategy. The leader needs to fine-tune his strategy with respect to the critical level of cheating. In certain cases the most economic manipulation strategy is to set the initial incidence of cheating just above the critical level, in order to 'prime the pump' of optimism. Information feedback then not only increases integrity amongst existing participants (as it did in Chapter 4) but also stimulates participation by people of increasingly greater sensitivity, who are in any case inclined to integrity whether or not other people cheat. This means that even in the case of trade, where greater integrity amongst existing participants will not occur, a virtuous cycle can still be exploited by inducing the selective entry of more honest traders. Because the initial crime rate is so close to the critical level, however, it is essential that the leader does not misjudge his manipulation strategy. If he fails to fine-tune manipulation, the outcome could be catastrophic.

As before, the simplest case to analyse is that of trade. Section 5.2 discusses the simultaneous determination of the crime rate and the participation rate for one-off trades. The changes induced by repeated trading are examined in Section 5.3. Sections 5.4 and 5.5 generalize these analyses to other cases, and Section 5.6 summarizes the main results.

5.2. The Incentive to Participate in Trade

It is assumed that an individual who does not participate is involved in no other encounters. Consideration of the possibility that he might quit to join another group is deferred until Chapter 13. It is also assumed that no participant has difficulty in making contact with a randomly selected partner no matter how small a proportion of the population participate in trade. This is reflected in the reward structure, which assumes that no special effort is required in order to participate in trade.

TABLE 5.1. *Trader's data set*

Strategy	Partner's strategy	
	Honesty	Cheating
Honest participation	a	$-b$
Dishonest participation	$a+b-g$	$-g$
Avoidance	0	0
Perceived probability	$1-p$	p

The simplicity of the two-stage decision problem is such that it can be collapsed into a one-stage decision between three alternative strategies—honest participation (0), dishonest participation (1), and avoidance (2). The rewards perceived by a representative trader are reported in Table 5.1. The condition for honesty to be preferred to cheating is exactly the same as before (compare (3.1) with $k=0$),

$$g \geqslant b. \tag{5.1}$$

What is novel is the set of participation conditions. Honesty is preferred to avoidance if

$$p \leqslant p^* = a/(a + b), \tag{5.2}$$

and avoidance is preferred to cheating if

$$g \geqslant (a + b)(1 - p). \tag{5.3}$$

It is worth noting that while condition (5.1), involving g, is independent of p, condition (5.2), involving p, is independent of g. It is only (5.3) that involves both. Since the conditions are linearly dependent, if any two of them are binding then the third is necessarily binding too.

Each of these conditions partitions $g - p$ space. Because of their linear dependence, the third partition passes through the intersection of the other two. Thus the three partitions generate only six regimes rather than nine, as shown in Fig. 5.1. Moreover, since pairs of adjacent regimes turn out to be associated with similar outcomes there are, in fact, only three fully connected areas that need to be considered.

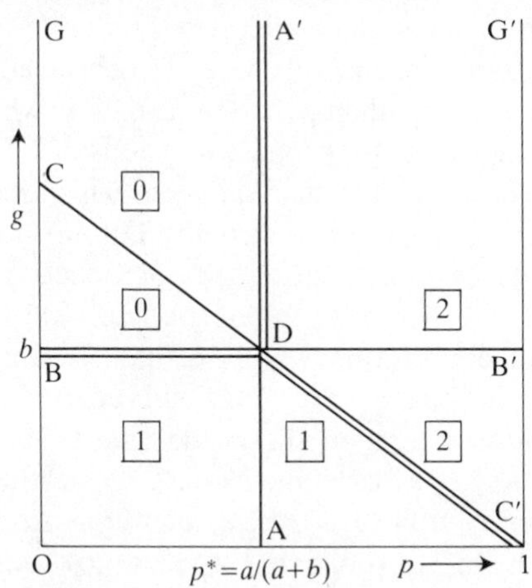

FIG. 5.1. Incentives with voluntary participation

The horizontal intercept OA of the vertical line AA′ represents the critical crime rate p^* at which honest participation and avoidance are equally attractive. The vertical intercept OB of the horizontal line BB′ represents the critical level of guilt at which honesty and cheating are equally attractive. The downward-sloping line CC′, which passes through the intersection D of AA′ and BB′ represents a state of indifference between cheating and avoidance. The downward slope indicates that the guilt required to deter cheating is greater at lower anticipated crime rates because potential cheats anticipate a good supply of suckers and so have less incentive to avoid trade altogether.

Because of the way in which weak, as opposed to strong, inequalities have been specified, indifference between honesty and any alternative leads to honesty being preferred. When honesty is not preferred, indifference between cheating and avoidance leads to avoidance being preferred. This is illustrated by the way that the boundaries are imputed to the various segments in Fig. 5.1. Points within and on the boundary of A′DBG involve honesty, points within and on the south-west boundary of OBDC′ involve cheating, whilst points within and on the south-east boundary of ADC′G′ involve avoidance.

The inequalities have been specified in this way for two reasons. First, analytical interest centres on honest participation (see below) and so it is convenient for the relevant region to include its boundaries (be closed). Secondly, for predictive purposes it is useful for the certainty of cheating to lead to a no-trade equilibrium, since it is no trade, rather than persistent cheating, which is observed in practice. This requires that someone who is indifferent between cheating and avoidance selects avoidance. This second condition means, however, that there is a discontinuity in the crime rate as an equilibrium of mutual cheating is approached, because at the limit the crime rate becomes indeterminate. To ensure that no trade can be an equilibrium when mutual cheating is not, it is assumed that when the no-trade situation is entered people use the last available information on the crime rate (namely $q = 1$) to decide whether to participate, and so never restart trade.

Apart from this complication, the conventions are useful because they make it possible to replicate the qualitative results obtained when there is a positive cost of participation (such as the cost of travel to market) without the algebraic difficulties which arise when such a cost is formally introduced into the analysis. (These difficulties stem chiefly from the fact that an altruistic leader is no longer indifferent between mutual cheating and mutual avoidance, so that additional terms enter into (5.5) below and make the relation between v and θ difficult to compute.)

Assume a uniform bivariate distribution of sensitivity and probability, as before:

$$F(s, p) = sp \qquad 0 \leqslant s, p \leqslant 1. \tag{5.4}$$

An altruistic materialistic leader measures the gains from trade within the group as

$$v = a(1 - x)(1 - q), \tag{5.5}$$

where a is the symmetric co-ordination gain, q is the crime rate, and x is the rate of avoidance, i.e. the proportion of the group membership that does not participate in trade.

An intensity of manipulation $\theta < b$ is pointless since no one will be honest. For $\theta \geqslant b$, the proportion of honest participants in the total group membership is

$$(1 - x)(1 - q) = [a/(a + b)][1 - (b/\theta)]. \tag{5.6}$$

Assume, as before, that manipulation costs are

$$c = \begin{cases} 0 & \theta = 0 \\ c_f + c_v\theta & \theta > 0. \end{cases} \tag{5.7}$$

Substituting (5.5)–(5.7) into the utility function

$$u = v - c, \tag{5.8}$$

and taking $\theta \geqslant b$ gives

$$u = [a^2/(a + b)][1 - (b/\theta)] - c_f - c_v\theta. \tag{5.9}$$

The first-order condition for a maximum is

$$a^2b/(a + b)\theta^2 = c_v, \tag{5.10}$$

whence

$$\theta^e = a[b/(a + b)c_v]^{\frac{1}{2}} \tag{5.11.1}$$

$$v^e = [a/(a + b)]\{a - [b(a + b)c_v]^{\frac{1}{2}}\} \tag{5.11.2}$$

$$u^e = [a/(a + b)]\{a - 2[b(a + b)c_v]^{\frac{1}{2}}\} - c_f. \tag{5.11.3}$$

It can also be deduced, using (5.6), (5.11.1), and (5.15) below that

$$q^e = (b + 2a)/(b + \{2a^2[b(a + b)c_v]^{\frac{1}{2}}\}) \tag{5.12.1}$$

$$x^e = [a/(a + b)]\{1 + 2(b/a)^2[(a + b)c_v/b]^{\frac{1}{2}}\}. \tag{5.12.2}$$

Overall, manipulation is worthwhile if $\Delta u - u^e - u(0) > 0$; since $u(0) = 0$, the condition for manipulation to be used is simply

$$c_f < [a/(a + b)]\{a - 2[b(a + b)c_v]^{\frac{1}{2}}\}. \tag{5.13}$$

It can be seen that the engineering of honest trading is much harder when participation is voluntary. It is not only that the fixed cost of manipulation must be recovered from a potentially smaller proportion of the population, but that the marginal return to manipulation is effectively reduced as well. This is because many of those whose guilt is increased will be pessimistic individuals who will merely reinforce their resolve to avoid participation altogether.

5.3. Recurrent Trading Encounters

The most dramatic implications of avoidance appear when repeated trade is considered. This is because a new equilibrium condition is introduced. The feedback of information on the actual crime rate must sustain the resolve of individuals to continue to participate in trade. Thus the actual crime rate must not exceed the critical crime rate p^*:

$$q \leqslant p^*. \tag{5.14}$$

This condition is relevant, of course, only if $p^* < 1$. It is readily deduced from the geometry of Fig. 5.1 that under a bivariate uniform distribution

$$q = (b + 2a)/[b + 2(a/b)\theta]. \tag{5.15}$$

The intensity of manipulation must therefore be sufficient to generate a crime rate from (5.15) which satisfies (5.14); this means that optimization with respect to θ is constrained by

$$\theta \geqslant \theta^* = [1 + (b/2a^2)(1 + 2a)]b > 0. \tag{5.16}$$

The leader's problem is again to maximize the utility function specified by (5.5)–(5.8). Recall from Section 4.4 that when participation in trade is obligatory, the solution of the recurrent trading problem coincides with the solution of the one-off problem because follower behaviour is independent of the expected crime rate and hence does not change in response to learning over time. With voluntary participation this no longer holds. A positive equilibrium level of manipulation must now exceed, not only the incentive to cheat, b, but the critical level θ^*. Three types of solution are now possible.

$$\theta = 0, \text{ no trade} \tag{5.17.1}$$

$$\theta = \theta^*, q^e = p^* \tag{5.17.2}$$

$$\theta > \theta^*, q^e = a[b/(a + b)c_v]^{\frac{1}{2}} < p^*. \tag{5.17.3}$$

With high costs of manipulation c_f, c_v, a high incentive to cheat b and low co-ordination gain a, the first solution—no manipulation—may be preferred. With somewhat lower costs, less incentive to cheat, or greater gains from co-ordination, it may become worth-while to manipulate up to the critical level. This will normally be appropriate when, in the absence of a participation constraint, a more modest but still positive level of manipulation would have been preferred. Finally, with low manipulation costs, a low incentive to cheat, and significant gains from co-ordination, intense manipulation in excess of the critical level may be preferred.

The situation is illustrated in Fig. 5.2. Suppose that initially participation is compulsory and that under these conditions the leader operates with a modest level of manipulation θ_0. This generates the horizontal response function Q_0Q_0'. The response function is horizontal because the follower's choice between honesty and cheating is independent of his expectations, so that any observed crime rate in the unit interval will generate the same outcome. Equilibrium is at E_0, the intersection of Q_0Q_0' and the diagonal OZ, and corresponds to a crime rate q_0.

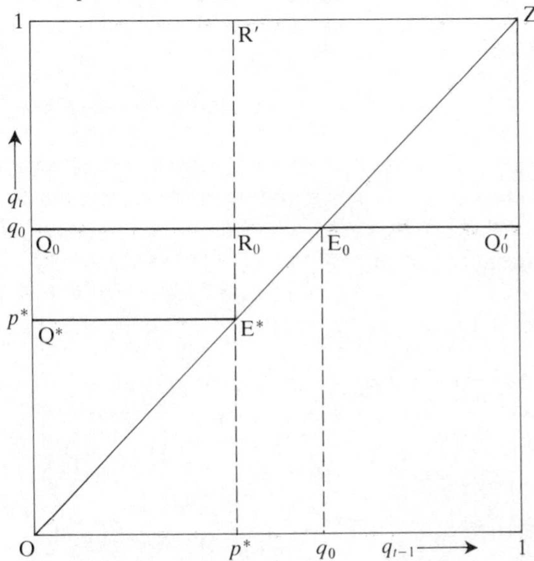

FIG. 5.2 Recurrent trading encounters

Now let participation become voluntary. The response function becomes discontinuous at R_0, corresponding to $q_{t-1} = p^*$. If $q_{t-1} > p^*$ the crime rate q_t becomes unity as all the honest traders quit. The response function is $Q_0R_0R'Z$ and the only intersection with the diagonal is at Z. As noted earlier, Z is, in fact, a no-trade equilibrium. Thus whilst all points except Z on the segment R'Z correspond, as indicated, to mutual cheating, Z itself does not. It is appropriate, however, to represent the no-trade equilibrium on the diagram by Z, if only because expectations of total pessimism reproduce themselves at Z, even if the observed crime rate does not.

The discontinuity means that the response function no longer intersects the diagonal at E_0 and so the economy will now converge on the no-trade equilibrium instead. To maintain trade when participation is voluntary, the leader must raise the intensity of manipulation to the critical level. This shifts down the left-hand portion of the response function to Q*E*, generating a new equilibrium at E*, where Q*E* meets OZ. By introducing discontinuity of response, therefore, voluntary participation makes leadership strategy under recurrent trading even more crucial. Failure to adjust the intensity of manipulation to quite small adverse changes in the environment (a decline in a or a rise in b, for example) can lead to the complete collapse of trade.

5.4. One-off Encounters in the General Case

The general case, in which guilt is probability-dependent, becomes quite complicated when participation is voluntary, but fortunately the main principles are easily discerned by diagrammatic analysis. The rewards anticipated by a typical follower are indicated in Table 5.2. This is, in fact, the same as Table 3.3, apart from the addition of a line of zeros for the non-participation option.

TABLE 5.2. *Follower's data set with voluntary participation: the general case*

Strategy	Partner's strategy	
	Honesty	Avoidance
Honesty (0)	h	$h - b - d$
Cheating (1)	$h + b - g$	$h - a - g$
Avoidance (2)	0	0
Perceived probability	$1 - p$	p

Follower behaviour is governed by three inequalities relating honesty to avoidance, avoidance to cheating, and honesty to cheating; they are respectively

$$p \leqslant p^* = h / (b + d) \tag{5.18.1}$$

$$g \geqslant (h + b) - (a + b)p \qquad (5.18.2)$$

$$g \geqslant b + kp, \qquad (5.18.3)$$

where, as before,

$$k = d - a. \qquad (5.19)$$

If (5.18.1) is satisfied, so that honesty is preferred to avoidance, then the crucial decision for behaviour is between honesty and cheating, and honesty is preferred only if (5.18.3) is satisfied as well. If, on the other hand, (5.18.1) is not satisfied, so that avoidance is preferred to honesty, then behaviour is determined by (5.18.2), and avoidance is chosen only if this is satisfied. It follows that if $p \leqslant p^*$ then behaviour is governed by (5.18.3) whilst if $p > p^*$ then behaviour is governed by (5.18.2).

If the gains from mutually honest encounters are very high relative to the loss associated with the being taken for a sucker then avoidance never occurs. Formally, if

$$h \geqslant b + d, \qquad (5.20)$$

then $p^* \geqslant 1$ and so the condition (5.18.1) is always satisfied. Conversely, if the gains from mutual honesty are negative then honest participation will not normally occur. Formally, if $b + d > 0$ then $h < 0$ implies $p^* < 1$ which implies that (5.18.1) is never satisfied. The restriction $b + d > 0$ merely rules out the unlikely case in which an honest person actually benefits by being cheated.

The conditions (5.18) divide $g - p$ space into three segments, as indicated in Table 5.3. A typical situation, in which $b > 0$, $d > a > h$, and $h < b + d$ is illustrated in Fig. 5.3. The schedule AA′ represents indifference between honesty and avoidance and is vertical because condition (5.18.1) is independent of g. The schedule BB′ represents indifference between cheating and avoidance and, from (5.18.2), slopes downwards whenever $b > -a$. The schedule intersects the p axis along the unit interval wherever $h + b > 0$ and $a > h$. The condition $b > -a$ is satisfied even if the incentive to cheat is negative, provided only that the *dis*incentive to cheat is no greater than the symmetric co-ordination gain. The condition $h + b > 0$ implies that cheating an honest victim is materially better than avoiding him, whilst $a > h$ means that mutual cheating produces a worse outcome than mutual avoidance. Since all these conditions are reasonable, it may be asserted that the form of BB′ illustrated in the figure represents the normal case.

The schedule CC′ represents indifference between honesty and

TABLE 5.3. *Strategic outcomes in* g – p *space*

Strategy	p condition	g condition
Honesty (0)	$p \leqslant h/(b + d)$	$g \geqslant b + kp$
Cheating (1)	$p \leqslant h/(b + d)$	$g < b + kp$
	$p > h/(b + d)$	$g < (h + b) - (a + b)p$
Avoidance (2)	$p > h/(b + d)$	$g \geqslant (h + b) - (a + b)p$

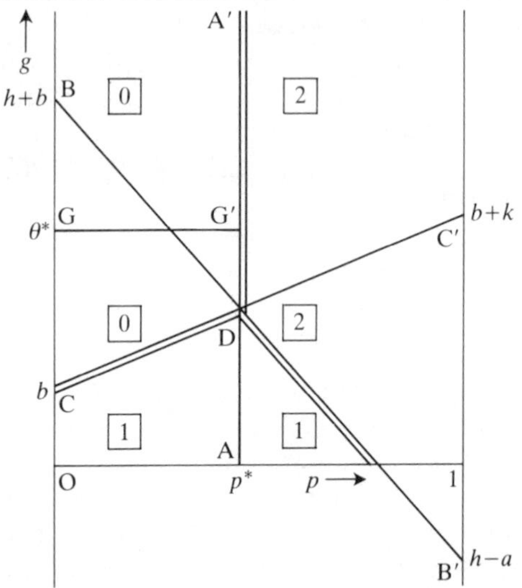

FIG. 5.3 Regimes in g – p space in the general case of voluntary participation

cheating, and is familiar from previous chapters. Comparison of the coefficients on p in (5.18.2) and (5.18.3) shows that this schedule normally cuts BB′ from below. The condition is that $h > 0$—in other words, that mutual honesty is better than mutual avoidance. As shown, CC′ is upward sloping but, as previous analysis has indicated, this need not be the case. The following discussion allows CC′ to be downward sloping too—but not so steeply as BB′.

The leader's objective in this general case is more complicated than in the previous example of trade. The leader is no longer indifferent whether those who do not participate honestly choose to cheat or to avoid trade. His objective varies with q and x separately and not just with the easily-computed product term $(1 - x)(1 - q)$. Algebraic expressions for q and x are extremely cumbersome in the general case, but using Fig. 5.3 we can write

$$q = A_1(\theta)/[A_0(\theta) + A_1(\theta)] \tag{5.21.1}$$

$$x = A_2(\theta)/\theta, \tag{5.21.2}$$

where $A_i(\theta)$ is the area in the figure associated with the ith strategy ($i = 0, 1, 2$). The areas are bounded from above by $g = \theta$, and so the As satisfy the equation

$$\sum_{i=0}^{2} A_i(\theta) = \theta. \tag{5.22}$$

The geometry indicates that, as in the previous example, the participation constraint will reduce both the average and marginal return to manipulation. It also shows, however, that the reduction in the marginal return depends on the probability-sensitivity of the choice between honesty and cheating, as measured by the slope k of the line CC′. If $k > 0$ then as the participation constraint tightens the least optimistic people, who require the most manipulation, decide not to participate instead of to cheat, and those that remain are relatively easy to manipulate. This improves the situation and prevents the marginal return from falling too quickly as participation falls. If, on the other hand, $k < 0$ then as participation falls the difficulty of manipulating those that remain increases, and the marginal return may fall very quickly as a result.

The properties of the line BB′ also affect the returns to manipulation, but in a more complicated way. Since BB′ is downward sloping, falling participation induces an accelerating number of people to switch from cheating to avoidance. If the leader prefers avoidance to cheating then this tends to keep up the marginal returns, whereas if he prefers cheating to avoidance it speeds up their decline.

5.5. Recurrent Encounters in the General Case

Chapter 4 presented a systematic analysis of the stability of recurrent encounters when participation was obligatory. Five main regimes were distinguished, and it is interesting to see how the stability of these regimes is affected when participation becomes voluntary. Recurrent encounters with voluntary participation are much easier to analyse than one-off encounters because probability perceptions are homogeneous rather than uniformly distributed, and as a consequence exact results can be readily obtained.

The results are illustrated in Fig. 5.4. Diagramatically, the significance of voluntary participation is that the response function acquires a discontinuity at the critical crime rate p^* given in (5.18.1).

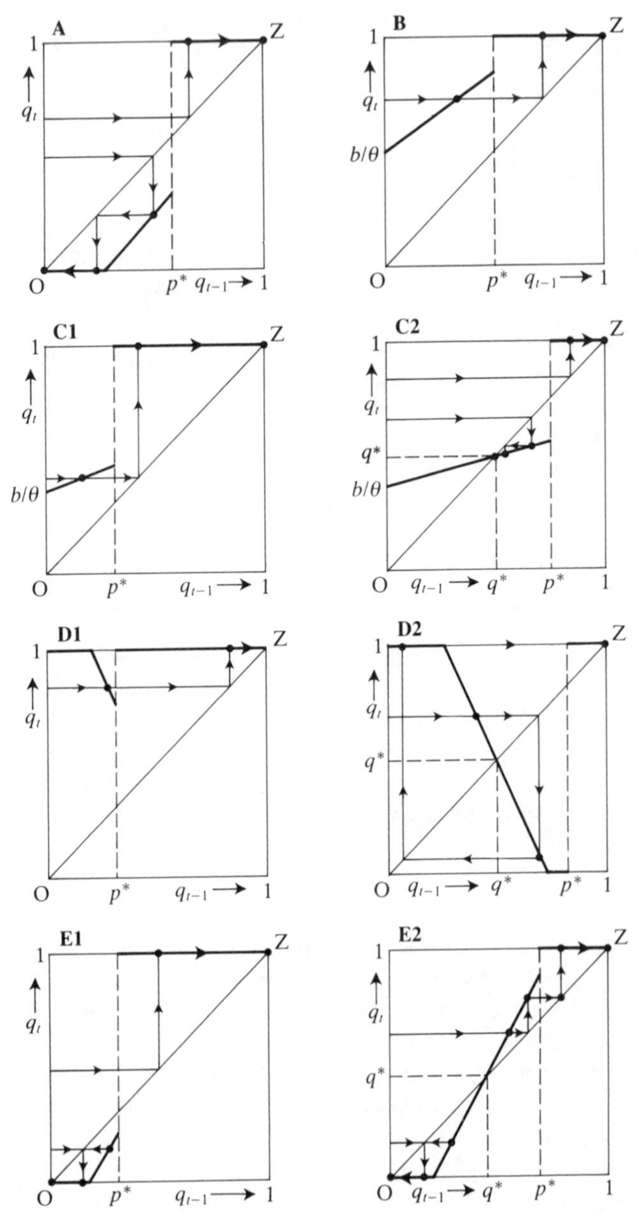

FIG. 5.4 Impact of voluntary participation on the equilibrium and stability of various regimes

A special case of this, involving a horizontal response function (and with $h = d = a$) was presented in Section 5.3. To the left of the discontinuity the response function is unchanged, but to the right it shifts up to run along the 'roof' of the diagram.

This discontinuity can eliminate certain equilibria altogether. Even if it leaves the equilibria unchanged, it can significantly destabilize the system (or add to instabilities already present). With compulsory participation regime A was basically a benign regime, converging spontaneously from all initial positions towards a totally honest equilibrium. But with voluntary participation, initial beliefs that people are dishonest will cause the most sensitive to avoid encounters, leaving only the least sensitive to participate. All of those who participate will then decide to cheat, and once everyone believes that everyone else will cheat no one participates at all. Voluntary participation turns initial pessimism into a major problem, and so makes announcements that engineer confidence crucial. The leader must either announce that everyone is honest at the outset, or manipulate guilt so that sufficient people are seen to have been honest in the first round. The latter strategy may require a higher degree of manipulation than is necessary to sustain the honest equilibrium once it has been achieved. Indeed no moral manipulation is required at all for this in regime A. The only valid reason for manipulation is to 'prime the pump' of confidence. Needless to say, in many cases the most effective way of doing this is not by manipulation but by announcement instead.

By contrast, in regime B, which is basically malign, avoidance has little consequence because the system will converge to a no-participation equilibrium anyway. The effect of voluntary participation is simply to hasten the process.

The remaining regimes C, D, and E, unlike A and B, all have an interior equilibrium crime rate. Regime E has other equilibria too. In regime C the interior equilibrium is stable, in D it is prone to oscillations, and in E it is unstable. In each case, however, the key question is the same: is the internal equilibrium crime rate greater than the critical value? If it is greater then the equilibrium is eliminated. If it is less than or equal to the critical value then the equilibrium remains, though its stability may be impaired. In regime C encounters collapse under a wide range of initial conditions. In regime D the oscillations no longer persist when out of equilibrium as the system converges eventually on a no-participation equilibrium. It is also possible in regime D that a minor change in the initial conditions could generate an immediate switch from totally honest participation to totally dishonest participation and thence to no participation at all (see quadrant D2). Regime E is highly unstable anyway, and so the consequences of voluntary participation, though qualitatively similar to those of D and E, are less noticeable.

The overall implications for the equilibrium and stability of recurrent trading are summarized in Table 5.4. The table confirms what the previous remarks suggest—that the tendency for encounters to degenerate by the escalation of cheating is far greater when participation is voluntary. Underlying this is the fact that the participation decision selects people according to their sensitivity. Voluntary participation allows the most sensitive people, who are

TABLE 5.4. *Equilibrium and stability of recurrent trading with voluntary participation*

Regime	Range of p^*	Equilibrium q-values		Effect of voluntary participation
		Stable	Unstable	
A	$p^* < 0$	1	—	$q = 1$ becomes an equilibrium
	$0 \leq p^* \leq 1$	0, 1	—	none
	$p^* > 1$	0	—	none
B	$-\infty < p^* < \infty$	1	—	none
C	$p^* < q^*$	1	—	$q = 1$ replaces $q = q^*$
	$q^* \leq p^* \leq 1$	q^*, 1	—	$q = 1$ becomes an equilibrium too
	$p^* > 1$	q^*	—	none
D	$p^* < q^*$	1	—	$q = 1$ replaces $q = q^*$
	$q^* \leq p^* \leq 1$	1	q^*	$q = 1$ becomes an equilibrium too
	$p^* > 1$	—	q^*	none
E	$p^* < 0$	1	—	$q = 0$ no longer an equilibrium
	$0 \leq p^* < q^*$	0, 1	—	$q = q^*$ no longer an equilibrium
	$p^* \geq q^*$	0, 1	q^*	none

Note: $q = 1$ signifies the non-participation equilibrium.

potentially the most honest, to respond to pessimistic expectations by avoiding trade altogether. The successful leader must not only prevent bad experiences from accumulating, but also promote good expectations by engineering the initial confidence that encourages the most sensitive to participate fully from the outset.

5.6. Summary

This chapter concludes the discussion of the moral manipulation of pairwise encounters by examining the interplay of participation and cheating in both one-off and recurrent situations. It shows that graphical techniques have a powerful role in analysing both the statics and the dynamics of behaviour.

The main conclusion is that with voluntary participation the leader's job becomes more difficult. Some of his effort in manipulation may be expended fruitlessly on individuals who have no intention of participating because they are pessimistic about the behaviour of other people. With recurrent encounters, a high incidence of cheating will discourage sensitive people, leading them to quit. This leads to an even higher incidence of cheating amongst those that remain. It is therefore important that the state of optimism is high. In some cases reputation effects can be exploited to begin with a modest initial degree of honesty, which builds up to greater honesty over time, whilst in other cases the only guarantee of satisfactory performace is a high level of manipulation from the outset.

Overall, the complications introduced by voluntary participation reinforce the view that group performance depends crucially on the quality of leadership. With unsophisticated leadership the benefits to the leader may not outweigh the costs, and if, as a result, the leader quits, then with no leadership at all the group may generate into total inactivity.

6

Reciprocity and Revenge

6.1. Alternative Patterns of Guilt

So far moral manipulation has been based on a very simple kind of rhetoric—namely, 'Cheating is wrong'. There are, however, at least three important questions that can be asked about this rhetoric, and each has potentially important consequences for followers' behaviour.

The first is whether cheating is really always wrong. Is it just as bad to cheat on someone who, it turns out, was planning to cheat on you as it is to cheat on someone who was honest? Many people take the view that what is wrong about cheating is taking advantage of the other party, and that advantage is only taken if the other party is honest. If the other party cheats then, in retrospect, it is only prudent that one cheated as well. This is one of the modifications of the moral system that is explored in this chapter.

Another way of justifying cheating on a cheat is to emphasize not only honesty but reciprocity as a virtue. In other words, the obligation is to reciprocate the honesty of the other party. Cheating on a cheat is not an offence—but merely reciprocity applied consistently to all the actions of the other party. The reason why cheating an honest person is wrong, on this view, is not that cheating is wrong, but that failure to reciprocate is wrong (L.C. Becker 1986).

The second issue is whether morality has to be about punishing crime rather than rewarding good behaviour. Does it make a difference if morality makes honest people feel good rather than cheats feel bad? The answer is 'Yes and No'. If participation in encounters is compulsory then it makes no difference, because the *relative* reward to honesty is still the same. But if participation is voluntary then it does make a difference, because, on average, the reward to participation is increased. On balance, therefore, a morality based on positive feelings towards oneself is better than one based on negative ones. Notice that this is true both in material terms—because of the material benefits of higher participation—and in purely emotional terms—good feelings are nicer than bad ones.

The final issue concerns the attitude the leader should take towards 'primitive' emotions such as the desire for revenge. The whole basis of leadership, as described in this book, is the manipulation of followers' emotions. This manipulation is targeted mainly to arousing a moral sense. But where people are spontaneously aroused to anger by a sense of injustice, should the leader attempt to manipulate this too? If so, should he subdue it, or intensify it; or channel the emotion in some other direction? This chapter examines the issues with respect to the anger of someone who has been cheated at trade. In this context the anger is directed against the cheat. The same issue is raised again in Chapter 11, however, in a somewhat broader context where anger is aroused by a general sense of social injustice and may be directed against the leader himself. In the present context, it is argued that anger can actually benefit economic performance by deterring cheating, but that in some cases the natural intensity of anger may be too great and so the leader may need to moderate it.

6.2. Reciprocity: A Diagrammatic Analysis

Changing the ethic from a unilateral condemnation of cheating to an obligation only to reciprocate integrity has a significant influence on behaviour. For example, when the unilateral ethic is used to modify material rewards in a PD, the game is converted to one of Harmony instead. But where an ethic of reciprocation is concerned, the PD is turned into an Assurance game instead. This is because the ethic of reciprocity does nothing to alter the material incentive to cheat on a cheat. This in turn means that as the intensity of manipulation is increased under an ethic of reciprocity, beliefs about the other party's integrity carry increasing weight. Thus ethical behaviour becomes highly sensitive to the degree of optimism or pessimism in this group. For obvious reasons, this leads to a potential high degree of instability in the behaviour of the group where repeated encounters are concerned.

The ethic of reciprocity generates the modified set of emotional rewards illustrated in Table 6.1. The corresponding changes in the follower's information set are indicated in Table 6.2. Conditions (5.2)–(5.4) are modified as follows:

TABLE 6.1 *Modified emotional rewards*

Strategy	Partner's strategy	
	Honesty	Cheating
Honesty	0	0
Cheating	g	0

TABLE 6.2 *Modified follower's data set*

Strategy	Partner's strategy	
	Honesty	Cheating
Honesty (0)	h	$h - b - d$
Cheating (1)	$h + b - g$	$h - a$
Avoidance (2)	0	0
Perceived probability	$1 - p$	p

Honesty is chosen instead of avoidance if

$$p \leqslant p^* = h/(b + d). \tag{6.1.1}$$

Avoidance is chosen instead of cheating if

$$g \geqslant (a + b) + [(h-a)/(1 - p)]. \tag{6.1.2}$$

Honesty if chosen instead of cheating if

$$g \geqslant -k + [(b + k)/(1 - p)], \tag{6.1.3}$$

where, following (3.2),

$$k = d - a. \tag{6.2}$$

The choice between honesty and avoidance remains unchanged because it was independent of guilt factors in the first place. It is still only optimism that affects the participation decision of the honest follower. The choice between avoidance and cheating is altered, however. The level of guilt required to induce avoidance is now higher by a factor $1/(1-p)$. This is unfortunate, because it means that with a given intensity of manipulation people are now more inclined than before to participate in order to cheat. The third condition, involving the choice between honesty and cheating, is affected in the same way. The level of guilt required to sustain integrity is increased by the same factor. One implication of this is that the intensity of manipulation required to sustain honest trading is no longer independent of expectations of cheating. The probability-independence of guilt among transactors, which previously required that $d = a$, now requires that $b + d = a$, which is much less likely to be satisfied in practice. The modified ethic raises the intensity of manipulation required to sustain honest trade, and introduces an element of instability into recurrent trading which was missing before.

Consider first a one-off encounter in the general case. Fig. 6.1 illustrates the typical case of probability-dependent incentives introduced in Section 5.4. The material incentive to cheat is higher if the other party is expected to cheat. It is assumed that, as before, $b > 0$, $d > a > h$, and $h < b + d$.

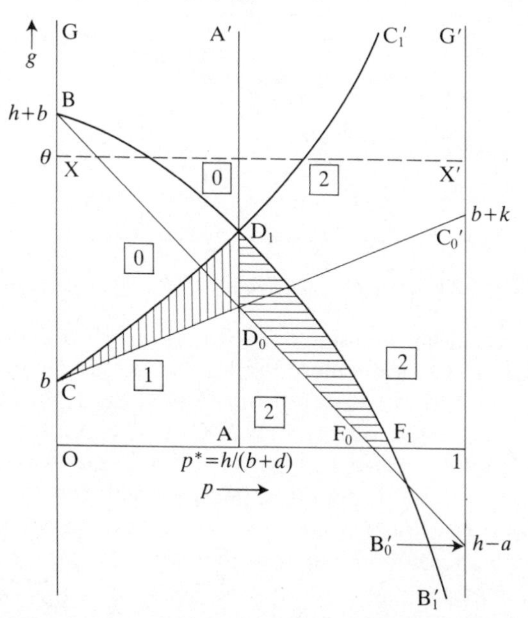

FIG. 6.1 Honesty, cheating, and avoidance with an ethic of reciprocity

The figure is adapted from Fig. 5.3. The original schedules are subscripted zero and the new schedules subscripted unity. Points which remain unchanged carry no subscripts at all. Only schedule AA′ remains unchanged. The schedule BB_1' lies above the original schedule BB_0', and the schedule CC_1' similarly lies above the schedule CC_0', indicating that for a given level of manipulation cheating is increased at the expense of both honesty and avoidance. The portion of the group switching from honesty to cheating is represented by the vertically-hatched area CD_0D^1, and the portion switching from avoidance to cheating by the horizontally-hatched area $D_0D_1F_1F_0$ (including the boundary segments CD_0 and D_0D_1). Both the participation rate and the crime rate are increased as more people join in and start to cheat.

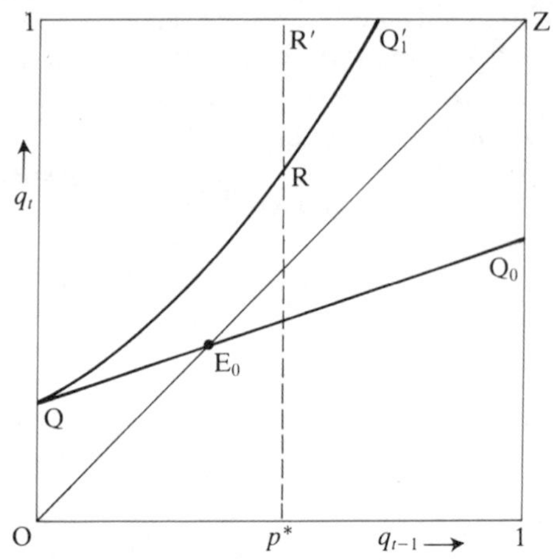

FIG. 6.2 Recurrent encounters with an ethic of reciprocity

The implications are no better so far as recurrent encounters are concerned. These are explored in Fig. 6.2. Consider first the case where participation is compulsory. Given the level of manipulation θ represented by the line XX′ in Fig. 6.1, the response function is $QQ_1'Z$. The portion QQ_1' corresponds to the schedule CC_1' in Fig. 6.1, with height scaled down by the factor θ. The kinked and curvilinear response function may be compared with the original straightline response function QQ_0' derived from the schedule CC_0'.

As shown, QQ_0' intersects the diagonal at E_0, giving the stable internal equilibrium crime rate q_0^*. By contrast the new response function $QQ_1'Z$ intersects the diagonal only at Z, indicating that the only equilibrium is one in which everyone cheats everyone else and in which, in consequence, no encounters will occur. Thus even with obligatory participation the dynamics of the system can be seriously affected by the change in the ethic.

When participation is voluntary a discontinuity of response is introduced at the critical probability p^*. The new response function QRR′Z, although discontinuous, has broadly similar properties to the old one QQ'_1 Z, but adjustment to Z is more immediate once the actual crime rate exceeds p^*.

The non-linearity of response engendered by reciprocity can lead to further complications, such as multiple internal equilibria. For example, with $k = 0$ and $\theta > 4b$ the segment QQ'_1 of the response function will intersect the diagonal twice (see Fig. 6.3). The lower of the two equilibrium crime rates q_1^* corresponds to a locally stable equilibrium E_1, whilst the higher one, q_2^*, corresponding to E_2, is locally unstable. The arrows in the figure indicate the direction of the adjustment trajectory in disequilibrium.

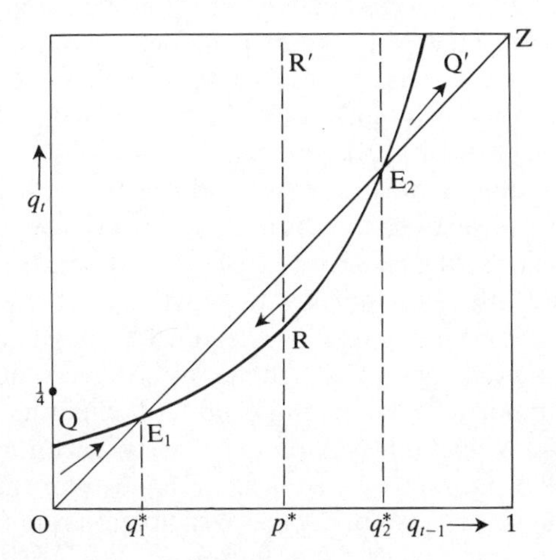

FIG. 6.3 Multiple internal equilibria

With voluntary participation one or both of these equilibria may be eliminated. In the case shown in the figure the response function acquires a discontinuity RR′ and as a result the higher, unstable equilibrium disappears. An initial crime rate in the region $p^* < q_1 < q_2^*$ which before would have induced a stable equilibrium at q_1^*, now leads instead to a no-encounter equilibrium at Z.

Finally, it is possible that the response function may be tangential to the diagonal. This unlikely situation generates an internal equilibrium crime rate which is stable to reductions in the crime rate, but unstable to increases. This highlights the proliferation of potentially unstable situations that the ethic of reciprocity can sustain.

6.3. Feelings of Satisfaction

The principle of reciprocity suggests that an outcome of mutual honesty may be regarded more positively than simply the absence of guilt. Reciprocation may involve not so much the avoidance of an emotional penalty as the conferment of a positive emotional reward. In practice, positive emotional satisfaction seems to be particularly common in groups where everyone believes that the others are 'pulling their weight' in pursuing the common ideal.

Positive emotional reward need not be linked exclusively to reciprocity, however. For example, the ethic that condemns cheating unilaterally can also offer positive emotional rewards if it encourages a sense of elation at honest behaviour. Thus while the Puritan ethic is often associated with guilt penalties, the same ethic, by emphasizing the virtue of self-discipline, can create an emotional benefit for the honest individual—a benefit which is sometimes perceived as pride or hypocrisy by others.

The distinction between positive and negative emotional rewards is of little significance so long as participation is compulsory. When the individual faces a binary choice between honesty and cheating, it is only the difference between the emotional rewards that counts, and avoiding a negative reward is equivalent to gaining a positive one. When participation is voluntary, however, the distinction becomes important. Self-punishment of cheating discourages participation while self-reward of honesty encourages it. Both reduce the crime rate by encouraging honesty at the expense of cheating but self-reward raises the participation rate too.

TABLE 6.3. *Follower's information set with feelings of satisfaction from reciprocity*

Strategy	Partner's strategy	
	Honesty	Cheating
Honesty	$h + f$	$h - b - d$
Cheating	$h + b$	$h - a$
Avoidance (2)	0	0
Perceived probability	$1 - p$	p

Suppose therefore that the emotional reward consists purely of a positive feeling of satisfaction, $f \geqslant 0$, associated with a mutually honest encounter. The total rewards perceived by a follower are indicated in Table 6.3. The relevant conditions governing follower behaviour are: honesty is chosen instead of avoidance if

$$f \geqslant -(b + d) + [(b + d - h)/(1 - p)], \qquad (6.3.1)$$

avoidance is chosen instead of cheating if

$$p \geqslant p^{**} = (h + b)/(a + b), \qquad (6.3.2)$$

and honesty is chosen instead of cheating if

$$f \geqslant -k + [(b + k)/(1 - p)]. \qquad (6.3.3)$$

The first point to note is that condition (6.3.3) is unchanged from (6.1.3) once f is substituted for g. This confirms that when participation is compulsory, feelings of guilt and feelings of satisfaction yield the same result. The next point to note is that the forms of the other two conditions are interchanged. It is now the choice between avoidance and cheating, rather than that between honesty and avoidance, that is governed by a critical probability that depends only on material rewards, independently of emotions. Conversely the choice between honesty and avoidance now depends upon emotions as well as material factors.

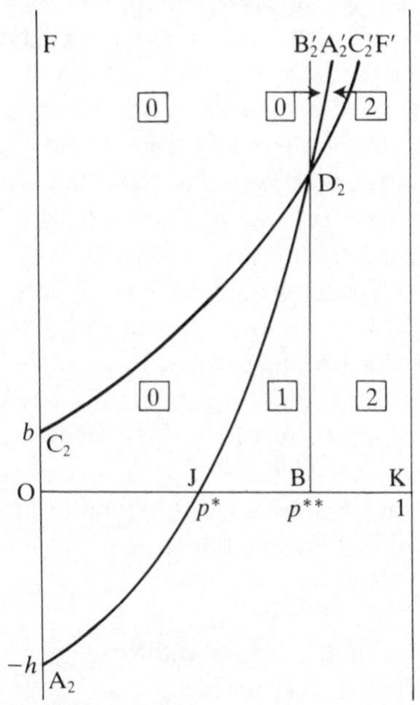

FIG. 6.4 Honesty, cheating, and avoidance with a positive feeling of satisfaction generated by reciprocity

Figure 6.4 reworks the example from the previous section, showing how the mechanics of avoidance are altered by substituting a feeling of satisfaction for a feeling of guilt. The schedule C_2C_2', indicating indifference between honesty and cheating, coincides with the schedule C_1C_1' in Fig. 6.1. The curvilinear schedule A_2A_2' replaces the vertical schedule AA′, although it should be noted that it cuts the horizontal axis at the same value p^*. The vertical schedule BB_2' replaces the curvilinear schedule BB_1'; BB_2' intersects the horizontal axis at a value $p^{**} > p^*$. Honest participants lie in the area $FC_2D_2A_2'$ (including the boundaries), cheats operate in the area OC_2D_2B (excluding the boundary C_2D_2B), whilst those who avoid encounters are found in the area $A_2'D_2BKF'$.

The figure indicates that with positive rewards the advantages of very intense manipulation are much greater. Because the critical level of participation is higher, at $p^{**} > p^*$, low levels of manipulation result in much more cheating, and in this example mutual cheating is, by assumption, worse than mutual avoidance (i.e. $a > h$). However, because participation is high, marginal returns to manipulation are high. Moreover, the returns to manipulation do not diminish as quickly as they do with guilt, because as manipulation increases the intensified feelings of satisfaction lead to still higher levels of participation as even the most pessimistic people are encouraged to join in.

Given the obvious benefits of engineering satisfaction rather than guilt, it may be asked why this approach is not more common in practice. To some extent, the engineering of satisfaction may be more widespread than it seems—because it typically occurs in small, tightly knit groups it may be less evident to outsiders than the more conspicuous engineering of guilt through education, religion, and media propaganda in large-scale groups. Another explanation may be that people are inherently less sensitive to positive feelings than they are to negative ones. Thus the cost of positive manipulation may exceed the cost of negative manipulation, and this higher cost may outweigh the higher returns obtained. This issue is considered further in Chapter 9.

6.4. Anger and Revenge

One of the most striking aspects of people's emotional constitution is anger. A quick-tempered individual is usually one of the most feared.

Anger is an interesting emotion because of all the emotions it appears to be the one most in conflict with common-sense ideas of rationality. As we shall see, however, this appearance is to some extent deceptive.

Anger often provokes people to do things which are damaging to their own material interests. It is essentially destructive: it typically damages the material interests of other people too. Anger is typically manifest in a state of arousal. When people have been aroused, a display of anger typically gives emotional release. This release allows the individual to obtain a net benefit from an action which is disadvantageous in purely material terms. The activity may therefore be rational once the individual has been aroused, because then the emotional reward outweighs the material cost involved.

Arousal is often a consequence of a display of anger by other people. Anger is frequently a conspicuous social phenomenon, and a display of anger by one person—particularly if directed against another—can prove contagious. This contagion means that individuals who commit themselves to materially self-damaging acts can often rely on others joining in with support if they do it conspicuously enough. This social user of anger is considered further in Chapter 11, where it is seen to be useful in deposing a leader who has lost the followers' confidence but is still trying to cling to power.

A major factor arousing people to anger appears to be a sense of injustice. More particularly, a belief that someone has betrayed a trust, and taken advantage of one's weakness and naïvety, appears to be crucial. In this context, the threat implicit in the capacity for anger is a potential source of benefit. It can act as a powerful deterrent to people who might otherwise have taken advantage in the first place. Anger improves the credibility of a threat to punish people, since they know that their injustice is liable to arouse the victim to punish them even at a material cost to himself. The fact that people cannot help being aroused is crucial in sustaining this credibility.

The anger mechanism is important in co-ordinating encounters such as trade. In Chapter 5 it was suggested that encounters are typically two-stage affairs, in which the first stage involves a decision whether to participate or not. In fact, it is probably more appropriate to think of an encounter as a three-stage affair in which the final stage involves a decision by a victim whether to pursue a cheat (see Table 6.4). In the absence of anger such pursuit is not normally worthwhile—the cost of pursuit normally exceeds the value of the material compensation obtained. But the emotional benefit of taking revenge may encourage pursuit. If the offending partner can anticipate this then the perceived net reward of taking a partner for a sucker is considerably reduced. At the same time, the anticipated emotional release obtained by taking revenge on an offender reduces the perceived loss of being cheated.

TABLE 6.4. *An encounter as a three-stage game*

Stage 1	Stage 2	Partner's stage 2	Stage 3		Partner's stage 3	
Avoid						
		Honesty				
	Honesty			Vengeance		
		Cheating	Arousal			
				Resignation		
Participate						Vengeance
		Honesty	———	———	Arousal	
	Cheating					Resignation
		Cheating				

The incentive to revenge is illustrated in Table 6.5 and the consequences for overall behaviour are indicated in Table 6.6. Both the victim and the offender, it is assumed, know the reward structure shown in Table 6.5. Revenge generates a feeling of satisfaction for the victim, f_r, which exceeds the cost of catching and punishing the offender $c_{rv} > 0$. When revenge is chosen, the cost to the offender is $c_{ro} > 0$. Given the inevitability of revenge when $f_r > c_{rv}$ both victim and offender can incorporate the consequences in their calculation of the benefits of cheating. The modified reward structure perceived by a representative individual in the general case is shown in Table 6.6.

Since $b' < b$ the net effect of revenge is equivalent to lowering the value of b. In the special case where the net benefit from revenge is exactly equal to the material loss suffered by the offender, this is the only change that occurs. If there is a discrepancy between the two, the effect of revenge will show up in an altered value of d as well.

TABLE 6.5 *Rewards associated with revenge*

Strategy	Reward to	
	Victim	Offender
Revenge	$f_r - c_{rv}$	c_{ro}
No revenge	0	0

TABLE 6.6 *Modification of overall rewards when an honest victim is known to take revenge on a cheat*

Strategy	Partner's strategy	
	Honesty	Cheating
Honesty	h	$h - b' - d'$
Cheating	$h + b'$	$h - a$

Note: $b' = b - c_{ro}$,
$d' = d + c_{ro} + c_{rv} - f_r$.

A reduction in the value of b substantially reduces the amount of manipulation required. All the results for optimal manipulation reported in the preceding chapters indicate that both optimal manipulation and the crime rate are positively related to b, whilst the leader's utility is negatively related to b. Moreover a reduction in b favours manipulation rather than monitoring—indeed under the conditions assumed earlier it does not benefit monitoring at all. The anger mechanism therefore increases the effectiveness of moral manipulation compared to monitoring.

6.5. The Moral Limits to Vengeance

The discussion above implicitly assumes that the feeling of satisfaction f_r is independent of the sensitivity of the individual concerned. The satisfaction generated by revenge, it is assumed, is different from the satisfaction generated by mutual honesty in that it is a natural and primitive feeling which occurs even in the absence of manipulation. Satisfaction from mutual honesty is, by contrast, not a natural feeling but one induced in sensitive people by moral manipulation.

The distinction is not, of course, as sharp as this in practice, but nevertheless it points to an important difference between the role of emotions in the natural state and their role when manipulation is employed. The natural satisfaction generated by revenge can achieve a measure of co-ordination in the absence of manipulation by raising the anticipated cost of cheating an honest person to the point where it is no longer worthwhile. The result is an 'economy of fear' in which the fear of vengeance sustains honest behaviour. Since in practice the capacity to take revenge often depends on the relative physical strength of the two parties, the economy of fear can lead to an unjust situation where nobody cheats the strong but everybody cheats the weak.

The economy of fear, therefore, tends to be reasonably efficient but very unfair. This explains the apparent paradox that while satisfaction from vengeance reduces the incentive to cheat, moral teaching on vengeance often seeks to reduce this satisfaction rather than increase it. The aim of moral teaching is not, primarily, to increase the efficiency of the economy of fear, but to replace it with a quite different system—a system which is superior on grounds of distribution. This question of distributive justice is taken further in Chapter 11, and the general issue is considered further in Chapter 13.

6.6. Summary

This chapter has conducted a 'sensitivity analysis' by varying the assumptions about emotions and their manipulation made in Chapters 2–5. The results turn out to be quite sensitive to the assumptions, as might be expected.

When an ethic of reciprocity replaces an ethic of unilateral integrity, a given intensity of manipulation results in higher participation but also a higher crime rate. This is because fewer insensitive people are deterred from participation, so more join in and cheat. When repeated encounters occur, the response of the current crime rate to the previous crime rate may become both non-linear and discontinuous, generating the possibility of multiple internal equilibria, and an internal equilibrium which is stable in one direction and unstable in the other. Only a relatively high intensity of manipulation can avoid these problems. All of this means that leadership becomes more difficult when an ethic of reciprocity is involved.

The engineering of positive feelings of satisfaction rather than negative feelings of guilt has, on balance, a beneficial effect on performance, although there is a particular problem at low levels of manipulation. The problem is that anticipated guilt no longer deters the less sensitive people from participating in order to cheat. As the intensity of manipulation is increased, however, even the least sensitive will want to enjoy the feelings of satisfaction that honesty brings, and so the problem becomes less acute. There is also the beneficial effect that sensitive but pessimistic individuals become much more inclined to participate at high levels of manipulation under reciprocity than they would under the guilt mechanism.

The natural emotion of anger can directly assist the leader in his task of manipulation. For when potential cheats recognize that their victims may get angry, the credibility of threats of reprisal is increased. Even though these threats may be materially self-damaging to the victim, the emotional benefit conferred by revenge may make the execution of the threat a rational strategy. The cost of reprisals anticipated by the cheat can substitute, in whole or in part, for the self-inflicted punishment of guilt. Given the earlier assumptions that the costs of monitoring are fixed costs independent of the size of fine, this means that natural anger reduces the cost of manipulation relative to the cost of monitoring. One interesting implication of this is that a leader who is committed to a manipulation strategy is more likely to tolerate, or even encourage, outbursts of anger within the group, than is a leader committed to monitoring. The manipulator may therefore broaden his strategy to encourage a more emotional approach to group activity as a whole, provided displays of emotions are kept within reasonable bounds.

PART III

Co-ordinating Work-Groups

7

Team spirit

7.1. Introduction

The analysis in Part II was based on a relatively abstract model of an economy in which all co-ordination involves pairwise encounters, and all manipulation is undertaken by a single all-powerful leader. The next three chapters develop a sequence of models of greater realism, by progressively relaxing these assumptions.

This chapter is concerned with teams. Two-person teamwork has already been discussed as a special case of pairwise encounter. This chapter extends the earlier analysis to teams of any size. Team work is shown to be a special type of Assurance game: the material benefit from individual effort is greater if other members are working hard as well. The main complication in modelling large teams stems from the interaction between team size and the proportion of team members who slack. This interaction is, of course, quite trivial in teams of size two. There is also a need to specify exactly how team members' beliefs about the efforts of their colleagues is determined. In teams of more than two members there may be a dispersion of sensitivities across other members of the group. A team member may need to take account of this in estimating what proportion of his colleagues are likely to be working hard.

Section 7.2 distinguishes between the engineering technology used by the team, which describes how output varies with team size when everyone is working hard, and the effort technology, which describes how the incidence of slacking governs the degree of underperformance. It introduces a simple assumption about effort technology, namely that the performance of the team is governed by the performance of its weakest link. The least dedicated worker alone determines how well a team of given size performs.

Two sets of beliefs about other people's efforts are considered. Section 7.3 studies optimal leadership strategy when each team member believes that the efforts of all other members are perfectly correlated—either everyone else is dedicated, or everyone else slacks. Section 7.4 considers an alternative case in which the effort decisions of other people are believed to be statistically independent of one another. It is shown that manipulation encounters greater difficulties in the second case. This is because each worker is much more pessimistic about overall team performance than he is about the performance of any individual member of the team. This problem becomes more serious, the larger is the team.

It is a feature of Assurance games that performance can be improved by engineering optimism through suitable announcements. It turns out that announcement effects are particularly important in large teams because the effectiveness of moral manipulation is inhibited by the pessimism about overall team performance noted above. In large teams the announcement must be *very* optimistic, because only exceptional optimism about individual performance will translate into optimism about team performance too. In a small team, or a team where efforts are believed to be perfectly correlated, this problem does not arise. In this latter case announcements need only be modestly optimistic and, in the absence of announcements, manipulation does not need to be so intense.

Section 7.5 examines the monitoring of team members. It emphasizes that monitoring team output is administratively easier than monitoring individual inputs of effort. Provided the leader can be trusted to administer fines fairly, a monitoring system using fines based on overall team performance will dominate all alternatives. This applies so long as members are concerned only with expected rewards—i.e., they are risk neutral. If members are significantly risk averse, however, then they may object to fines based on overall team performance which they cannot themselves control. In this case, announcement may again become the optimal strategy.

The effort technology, it was indicated above, normally generates decreasing returns to team size. At the same time, the engineering technology may afford increasing returns. Section 7.6 examines the interaction between these opposing forces, and shows that often they are finely balanced. As a result, optimal team size may respond discontinuously to changes in the environment. The interaction between engineering technology and effort technology in teams of variable size provides a basis for a significant extension of the theory of the firm, and is considered further in Chapter 8.

7.2. Complementarity of Individual Efforts

The essence of teamwork is that the efforts of different team members are complementary. Some writers use the expression 'teamwork' to cover all types of group activity, but that is not the approach here. Teamwork is regarded as a special type of encounter. Encounters have so far been exemplified mainly by transactions, but transactions exhibit technological separability, which team production does not. In team production cheating—or slacking as it is called in this context—impairs the productivity of other team members, whereas in a transaction, default by one of the parties does not automatically cause their partner to default as well (Alchian and Demsetz 1972).

The technology of team production involves two distinct aspects, which are not always as sharply distinguished as they need to be. One—called here the *engineering technology*—determines how the maximum potential output is related to the number of members of the team. It is the engineering technology which is referred to when production is said to exhibit either increasing, constant, or decreasing returns to scale. Returns to scale, in the present context, are determined by whether the maximum attainable average product increases, is constant, or declines with respect to team size.

The second aspect is the *effort technology*. This is relevant because the potential described by the engineering technology is not always attained due to slacking by some of the members. The effort technology describes how underperformance is related to the number of members who slack. The analysis below is based on a very simple effort technology. It implies that dedicated efforts are strictly complementary, and is the formal analogue of the intuitive idea that a chain is only as strong as its weakest link. Thus the loss of output due to slacking is the same for just one slacker as it is for any number of slackers. The loss of output is measured in terms of average productivity (rather than total product): actual average productivity falls short of potential average productivity by the same fixed amount independently of the size of the team.

The size of team affects the engineering technology and the effort technology in quite different ways. The larger the size of team, the greater is the division of labour that can be implemented, in terms of the number of specialized roles that can be defined. A team member is an indivisible human resource, and in a large team each member can be kept fully utilized performing a narrowly defined task. The classic example of returns to specialization between strictly complementary members of a team is the division of labour in the pin factory mentioned by Adam Smith (1776)—the modern equivalent being the assembly line in a manufacturing plant or the 'disassembly' line in a raw material processing plant. Large size also permits fine-tuning in matching people to roles in accordance with their personal comparative advantage. Overall, therefore, it seems likely that, provided complementary non-labour inputs are in elastic supply, the engineering technology will afford increasing returns to scale.

The picture with effort technology is rather different, however, as the following analysis shows. Under certain conditions the risk of underperformance increases with the size of team. This is because a large team is more likely than a small team to contain a 'weak link'. Thus increasing returns in the engineering technology may be offset by decreasing returns in the effort technology, as a result of which the optimal size of team may be relatively small.

7.3. Manipulating Team Effort

In terms of the typology of encounters presented in Chapter 3, teamwork is interesting because it exemplifies regime IV. This regime sustains spontaneous dedication provided members are confident that all other members are dedicated too. This is in sharp contrast with transactions, which exemplify regime III, in which no one is honest (dedicated) unless they are either monitored or manipulated. The key to this difference is that with team production the best response to other people's dedication is dedication whilst in a transaction the best response to other people's honesty is to cheat. This suggests a prima facie case that manipulation is less important in teamwork than in trade, but that suitable announcements regarding the integrity of others will carry much more weight.

The leader of the group simultaneously co-ordinates all the teams. The case where each team has its own leader is considered in the next chapter. The leader of the group chooses between manipulation and monitoring, as before. There is one special twist, however, which is that the leader can also choose the size of team. In the trade case all encounters were assumed to be bilateral.

In the context of team production, realism demands that multilateral co-ordination be considered as well.

Suppose, then, that there is an altruistic leader who allows members of a team to share all the output between themselves. The team produces a private good (public goods are considered in Chapter 10) which is distributed equally between its members. Under manipulation this is the end of the story, but under monitoring those who slack are liable to a fine. This fine is retained by the leader and not distributed to other members of the group.

When participation is compulsory, each member faces a single choice between dedication and slacking. If everyone is dedicated then everyone receives a portion $y > 0$ of the maximum potential output, but incurs an effort penalty $e > 0$. If one member slacks whilst the others are dedicated then everyone receives a portion of output $y - \Delta y$, where $\Delta y (0 < \Delta y < y)$ is the marginal productivity of effort when everyone else in dedicated. The member who slacks avoids any cost of effort, but the others do not. It is assumed that (unlike Chapter 2) effort is worthwhile from a broadly materialistic point of view, i.e. $\Delta y > e$.

Because of the 'weakest link' effort technology, the outcome from either dedication or slacking depends only upon whether any one (or more) of the other members slack. Given his belief about the incidence of slacking at the individual level, the team member must infer the consequences of this at the group level.

The simplest pattern of inference arises when the slacking amongst other members is believed to be perfectly correlated. This belief is trivially true for a two-person team. Given that, by assumption, all individuals face the same material incentives, it could also be true for a larger team as well. Two further conditions must normally be met, however. The first is that there is no manipulation—because if manipulation occurs differences can arise because of differences in sensitivity within the team. The second is that everyone should have similar beliefs about other people's behaviour. This second condition is satisfied in a one-off encounter if the leader has made an announcement, and it is automatically satisfied in recurrent encounters because, by assumption, everyone's expectations are based on the same past experience. It is assumed below that individuals believe in perfect correlation even when these conditions are not fully satisfied.

If this assumption is accepted then the probability P that one or more of the other members slacks is equal to the probability p that any given member slacks:

$$P = p. \tag{7.1}$$

An alternative approach, based on the statistical independence of individual decisions, is considered in Section 7.4.

TABLE 7.1. *Member's data set for team production*

Strategy	All partners dedicated	Some partners slack
Dedication	$y - e$	$y - \Delta y - e$
Slacking	$y - \Delta y - g$	$y - \Delta y - g$
Perceived probability	$1 - P$	P

The reward structure for team production is illustrated in Table 7.1. It correponds to a special case of Table 3.3, where

$$a = \Delta y - e > 0 \tag{7.2.1}$$

$$b = \mathrm{e} - \Delta y < 0 \tag{7.2.2}$$

$$d = 2\Delta y - e > 0 \tag{7.2.3}$$

$$h = y - e > 0. \tag{7.2.4}$$

The intensity of guilt required to discourage slacking is given by the familiar-looking inequality

$$g \geqslant b + kP, \tag{7.3.1.}$$

where

$$k = d - a = \Delta y. \tag{7.3.2}$$

Since $b < 0$, $k > 0$, and $b + k - e > 0$, regimes IV or VIII apply (see Table 3.5). Since VIII affords constant returns to manipulation, an interior maximum for one-off manipulation cannot occur there. Attention must therefore be focused on regime IV. This is the regime most characteristic of an Assurance game.

A narrowly materialistic leader ignores effort costs and so measures reward by

$$y_1 = y - q\Delta y. \tag{7.4}$$

He maximizes utility subject to the familiarcost function and the crime-rate function for regime IV indicated in Table 3.5. The optimal intensity of manipulation and the implied incidence of slacking are

$$\theta_1^e = e/(2c_v)^{\frac{1}{2}} \quad (7.5.1)$$

$$q_1^e = e(c_v/2)^{\frac{1}{2}}/\Delta y, \quad (7.5.2)$$

where c_v is the marginal cost of manipulation. The leader obtains a reward

$$v_1^e = y - e(c_v/2)^{\frac{1}{2}}, \quad (7.5.3)$$

which, net of manipulation costs, gives utility

$$u_1^e = y - e(2c_v)^{\frac{1}{2}} - c_f. \quad (7.5.4)$$

These results are surprisingly simple because, given the crime-rate function, the loss of leader's utility due to slacking is independent of the marginal productivity of effort. For reasons already indicated, the loss is independent of the size of team as well. Manipulation varies directly with the effort penalty and inversely with the square root of the marginal cost of manipulation. The crime rate does, however, depend on marginal productivity as well as the effort penalty, as one would expect.

Given the crucial role of effort in team production, the welfare implications are better understood in terms of broad materialism, however. The broadly materialistic leader will, as expected, use a lower intensity of manipulation and so tolerate a higher incidence of slacking. The results may be obtained simply by substituting (7.2.1), (7.2.2), and (7.3.2) into the expressions for θ^e and q^e in Table 3.6, to get

$$\theta_2^e = e[(1 - (e/\Delta y))/2c_v]^{\frac{1}{2}} \quad (7.6.1)$$

$$q_2^e = (e/\Delta y)\{c_v/2[1 - (e/\Delta y)]\}^{\frac{1}{2}} \quad (7.6.2)$$

$$v_2^e = y - c_f - \{[(1 - (e/\Delta y)]c_v/2\}^{\frac{1}{2}} \quad (7.6.3)$$

$$u_2^e = y - c_f - e\{2[1 - (e/\Delta y)]c_v\}^{\frac{1}{2}} \quad (7.6.4)$$

In practice, however, manipulation strategy is likely to be dominated by an announcement strategy (assuming that the cost of an announcement is reasonably small). Simply by announcing that everyone else will be honest, the leader can induce everyone to be honest. The announcement therefore has the happy property of being self-validating. What is more, the pattern of behaviour it induces is consistent with members' beliefs that the decisions of other members are completely correlated.

It is, in fact, unnecessary to make an announcement that everyone else will be dedicated. It is sufficient to announce that the probability of a weak link occurring is less than

$$P^* = -b/k = 1 - (e/\Delta y). \tag{7.7}$$

This announcement is, in fact, unnecessairly pessimistic, for it will avoid any weak link altogether. But if members of the team are initially very sceptical, the pessimism may give the announcement greater credibility.

7.4. Independent Slacking Decisions

An alternative assumption about probability beliefs is that members believe that other people's decisions whether to slack are statistically independent. Let $m \geqslant 2$ be the size of team (the size of team, m, is distinct from the size of group, n, because many teams may belong to the same group). If the probability that any given member slacks, p, is independent of whether any other member slacks then the probability P that at least one of the other $m - 1$ members slacks is

$$P = 1 - (1 - p)^{m-1} \tag{7.8}$$

Equation (7.8) is graphed for various values of m in Fig. 7.1.

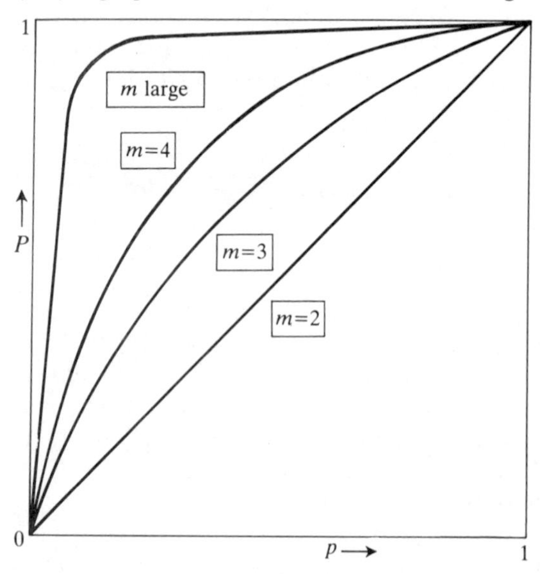

FIG. 7.1 Relationship between individual slacking and the appearance of a 'weak link' amongst other members of a team, in probabilistic terms

The basic message of the preceding section—that announcement is more effective than moral manipulation—still applies, and indeed applies with even greater force. The main qualification is that the announcement must claim an even higher degree of integrity as team size increases. As size approaches infinity, only an announcement that everyone else is certain to be dedicated will suffice. Except for this neighbourhood around $p = 0$, the material incentive to cheat becomes increasingly independent of probability; even if a given individual is thought unlikely to cheat, large team size makes it practically certain that someone will cheat. Thus outside the neighbourhood of total confidence, the problem becomes increasingly similiar to the one encountered in the case of trade, where the incentive to cheat is independent of probability beliefs.

So far as a manipulative strategy is concerned, the optimal intensity of manipulation is increased by independent probabilities, and the optimal incidence of slacking rises as well. The magnitude of the changes increases, as expected, with the size of team.

If the search for an optimum is confined to the region of diminishing returns, where $\theta > e$ then the crime rate is given by

$$q = b(m)/\theta, \tag{7.9}$$

where

$$\begin{aligned} b(m) &= \int_{1-(e/\Delta y)^{1/(m-1)}}^{1} [e - \Delta y(1-p)^{m-1}]\mathrm{d}p \\ &= [1-(1/m)](e^m/\Delta y)^{1/(m-1)}, \end{aligned} \tag{7.10}$$

so that, for large m, the economy approximates regime III:

$$\lim_{m \to \infty} b(m) = e. \tag{7.11}$$

Substituting (7.9) and (7.10) into the narrowly materialistic leader's utility function (7.4) in the usual way, taking $y \geqslant 2\Delta y$, and solving the first-order conditions gives

$$\theta_1^e = \{[1-(1/m)]\,(e^m\,\Delta y^{m-2})^{1/(m-1)}/c_v\}^{\frac{1}{2}} \tag{7.12.1}$$

$$q_1^e = \{[1-(1/m)]\,(e/\Delta y)^{m/(m-1)}c_v\}^{\frac{1}{2}} \tag{7.12.2}$$

$$v_1^e = y - \{[1-(1/m)]\,(e^m\,\Delta y^{m-2})^{1/(m-1)}c_v\}^{\frac{1}{2}} \tag{7.12.3}$$

$$u_1^e = y - 2\,\{[1-(1/m)](e^m\Delta y^{m-2})^{1/(m-1)}c_v\}^{\frac{1}{2}} \tag{7.12.4}$$

Results for a broadly materialistic leader can be derived in an analogous way, but no new insights are revealed. By plotting these results for trial values of $e, \Delta y$, and c_v, and considering the limit as m becomes large, it can be confirmed that diseconomies of manipulation are incurred with increasing size of team.

7.5. Team-Based Incentives

One of the features of team technology is that output has to be monitored if there is to be any systematic basis for sharing the product. This means that output must be monitored even under manipulation. It also means that even when inputs are not monitored, a cheap monitoring strategy can be employed that relates individual rewards to team output.

The strategy is to impose heavy fines on all the members of a team that fails to achieve its potential output. If participation in team production is compulsory, then a large but finite fine will always suffice to induce effort, provided no member believes that some other member is absolutely certain to cheat. Under these circumstances team-based incentives will always be preferred to manipulation since they eliminate slacking without any further expense.

Such incentives have more limited value when participation is voluntary, however. The large size of the fine required (particularly in a large team) means that this draconian approach will discourage the pessimistic from joining the team. To raise the expected value of participation, the sharing rule can, however, be modified to pay out the accumulated surplus from previous fines as a bonus for realizing the potential. Under this arrangement the size of the fine is dramatically reduced, as the following calculations show.

With a probability of team default P, the average repayment that will have accumulated for distribution as bonus is $[P/(1-P)]f$ where $f > 0$ is the size of the fine. The probability that a bonus will be paid is $1 - P$. It can be deduced that everyone will be dedicated if

$$f \geqslant b + kP. \qquad (7.13)$$

This constraint is satisfied by a fine as low as $f = e < \Delta y$. If $y \geqslant 2\Delta y$ then the fine can be held as a natural hostage; team members receive nothing until the leader has monitored team output, and if slacking has occurred then each member receives only $y - \Delta y - f \geqslant 0$ instead of $y - \Delta y$ as he would have done under manipulation. If everyone has been dedicated then everyone receives, on average, $y + [P/(1 - P)] f$ instead.

On this basis, team incentives completely dominate manipulation. They dominate the monitoring of individual effort too. In practice, of course, there are complications—notably the additional administrative costs associated with contingent payments, and the high degree of material risk to which team members are exposed as their rewards fluctuate over time. Risk-averse individuals may not wish to participate in team incentive schemes at all. Because of these problems it is doubtful if team incentives dominate announcement strategies. They both eliminate slacking, but announcement is normally cheaper to administer than a system of contingent payments.

Overall, this suggests that in the context of team production, the usual ranking of strategies will be

(1) announcement;
(2) team incentive;
(3) either moral manipulation or individual monitoring.

7.6. Optimal Size of Team

The leader of a large group has a choice between dividing up the group into many small teams or a few large ones. It is assumed that all teams have the same engineering and effort technologies and that a unique optimal size exists. The problem of whether the group will divide up neatly into an integer number of teams is ignored by allowing the number of teams to be a continuous variable. We do allow for the possibility, however, that the optimal size of team exceeds that of the group, so that the whole group consists of a single team of suboptimal size.

Announcement strategies and team incentives solve the effort problem in a manner which is independent of team size. So too does individual monitoring. There are two main reasons for this. First, it has been assumed that the fixed costs of these systems are related to the size of the group and not the size of the team, so that the replication of teams does not incur additional costs. Secondly, the marginal productivity of effort and the cost of effort are taken as independent of team size, so that the material incentive to cheat does not vary with size. In each of the cases mentioned above the team can therefore expand with constant per capita transaction cost. As a result, it is the engineering technology alone which determines the optimal size of team.

Under manipulation, the situation is more interesting—at least where independent probabilities are concerned (with perfectly correlated probabilities, transaction costs are again invariant to size). Given independent probabilities, manipulation is most effective, relative to the alternatives, when team size is small.

To calculate optimal team size, note that the narrowly materialistic leader maximizes per capita output net of direct transaction costs. The optimal size is derived simply by substituting the envelope technology function,

$$y = y(m), \tag{7.14}$$

into the leader's utility (7.12.3). The calculation of the first-order condition is, however, extremely difficult because of the way m enters as a power applied to other terms. Because the principle is so straightforward, however, it can be illustrated diagrammatically.

In Fig. 7.2 the schedule YY′ in the upper segment represents the engineering technology. Average product is plotted vertically and size of team horizontally, so that the upward slope represents continuously increasing returns to scale. The schedule TT′ represents the transaction costs incurred when moral manipulation is exercised within the constraints of the effort technology. Exactly half the transaction cost is an indirect cost associated with loss of potential output, and the other half is a direct cost associated with manipulation expenses. Leader's utility is represented by the schedule UU′, whose height is equal to the vertical discrepancy betweeen YY′ and TT′. To achieve a unique optimum size, it is necessary that the curvature of YY′ exceeds that of TT′, as illustrated in the figure.

This curvature condition ensures that in the lower quadrant the curve RR′, which is marginal to YY′, cuts the curve CC′, which is marginal to TT′, from above. The optimum size OM is determined by the intersection E of RR′ and CC′, which in turn corresponds to the peak M* in the quadrant above.

The introduction of an alternative technique may render manipulation obsolete. Suppose, for example, that a team incentive is introduced incurring a direct transaction cost, associated with the administration of contingent payments, measured by the height OZ of the horizontal line ZZ′. Efficient choice of strategy now implies operating on the minimum transaction cost envelope TXZ′.

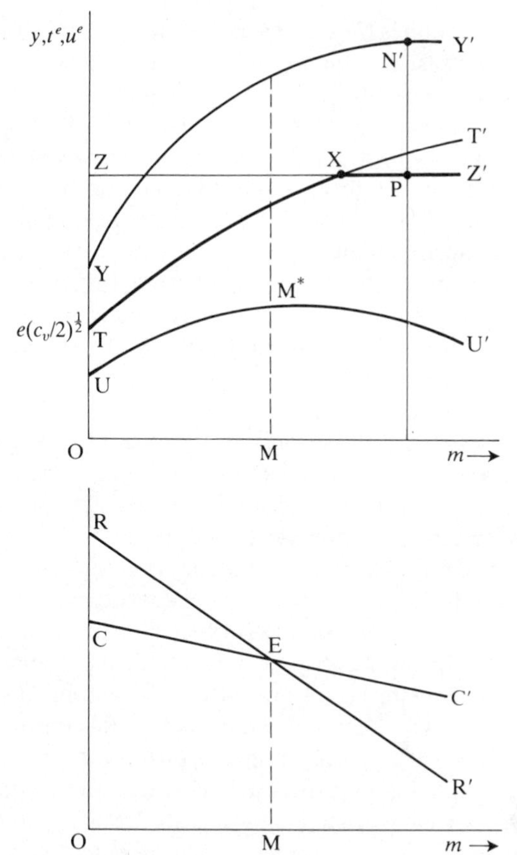

FIG. 7.2 Determination of optimal size of team

Note: The origin corresponds to $n = 2$ the schedule TT' is asymptotic to a horizontal line of height $(e\Delta yc_v)^{\frac{1}{2}}$

Comparing the height of TXZ′ with the height of YY′ shows that it is now optimal to introduce team incentives and expand team size to infinity. This strategy will, of course, encounter the limit set by group size, as indicated by the horizontal intercept ON. Leader utility is constrained to N′P as a result. But so long as N′P > M*M the expansion strategy is still worthwhile.

Because moral manipulation is the only strategy with size-dependent transaction costs, and these costs increase with size, it follows that as team size expands, leaders may switch from moral manipulation to other strategies but never the other way round.

Moral manipulation is thus essentially a 'small-team' strategy. This accords with casual empiricism, which suggests that moral manipulation is typically associated with small teams rather than large ones.

It is clear that on the whole the determination of team size involves a delicate balance of forces. A minor change in the environment—whether in the slope or curvature of the engineering technology function or in the relative transaction costs of different strategies—can precipitate a significant change of size. The implied discontinuity of response means that the comparative statics of team size do not provide unambiguous results of the kind that were derived for fixed team size earlier on.

Further complications arise if the potential size-dependency of marginal productivity and effort cost is taken into account. Critics of mass production and the factory system, for example, often allege that the greater specialization effected within large teams alienates members from their work, and so causes the psychological cost of effort to rise. The size-dependency of effort cost has become a key element in arguments that 'small is beautiful', and that large teams must be split into coalitions of smaller teams to enrich the jobs involved. Rising effort cost not only drives a greater wedge between broadly materialistic and narrowly materialistic objectives but, as one would expect, exacerbates the problem of manipulation too. Rising effort cost reinforces the argument advanced above that manipulation is relatively most effective in small teams. As team size expands, manipulation costs are likely to accelerate due to alienation whereas the costs of other strategies are likely (under the assumptions here) to remain unchanged. If alienation becomes translated into anger, however, instead of merely disaffected slacking, then the other strategies may be undermined as well. This idea is explored further in Chapter 11.

7.7. Summary

This chapter has examined the determinants of team performance within an economy where a single leader manipulates all teams. Team performance is governed by the interplay of (i) the marginal productivity of an individual worker who is added to the team to increase its size; (ii) the marginal productivity of effort, which is the increase in team output achieved if an individual worker, who would otherwise be the weakest link, decides to work hard; (iii) the cost of effort perceived by each worker; (iv) the structure of workers' beliefs about other people's effort; (v) the costs of manipulation; (vi) the cost of announcement; and (vii) the costs of monitoring—in particular the cost of monitoring team output and implementing an incentive scheme based on this.

Team performance is optimized both by choosing the appropriate size of team and by selecting the appropriate leadership strategy associated with a team of that size. The two choices are interdependent in the sense that the costs of alternative leadership strategies affect the optimal team size.

It was noted in Chapter 2 that individual effort within an economy will depend upon a mixture of leader-specific, group-specific, and situation-specific factors. This also applies to team effort, but with the complication that some of the additional factors—such as engineering technology—may be specific to all three. The best available technology may be affected by leader-specific trade secrets, group-specific skills, and situation-specific factors such as the climate and the availability of materials. Not all of the additional factors raise such complications, however. When team members are reasonably well informed, for example, their structure of beliefs about other people's effort when subjected to manipulation will be influenced by the distribution of moral sensitivity within the group, and this is clearly a group-specific phenomenon.

8

Intermediators: The Middle-Class Middlemen

8.1. Introduction

Leadership has been analysed, up to this point, on the assumption that it cannot be delegated. The role has been assumed to be essentially a unitary one. This is a very extreme assumption which needs to be relaxed in the interests of realism. Consider, for example, a national leader with responsibility for promoting internal trade. It is unrealistic to expect such a leader to rely exclusively on direct communication with every citizen. A good deal of his influence is likely to be mediated by an élite. The leader influences the members of this élite and these members in turn transmit his values to the citizens with whom they deal.

Among the most influential members of the élite will be those whose role it is to intermediate in trade. Intermediation reduces transaction costs for various reasons. It can perform a break-bulk function in which wholesale-size lots are broken up into smaller retail-size lots suitable for the final customer. It can provide a stock-holding service, allowing customers to inspect goods before taking delivery and to purchase them on demand instead of only to order. The most important function of intermediation in the present context, however, is to overcome the problems created by distrust.

An intermediator who gains a reputation for integrity can establish an indirect link between traders who cannot trade directly because they do not trust each other. If each trader recognizes that the intermediator has a reputation that he does not have, and believes that this reputation is warranted, then he will be happy to place a hostage with him. The most natural hostage is the product itself—the trader accepts that he must prepay for what he buys and accept post-payment for what he sells. Because the intermediator can successfully demand hostages, he is effectively insured against default by his customers, and so his distrust of them is no obstacle to trade.

An intermediator who demands hostages is effectively relying on a monitoring mechanism. Another possiblility is that the intermediator can reduce his risks by moral manipulation of his customers, so that they do not cheat on him even though they might be inclined to cheat on other people. The intermediator therefore faces a choice between monitoring and manipulation, exactly as the national leader does.

The role of intermediators in promoting trade makes it clear that leadership is indeed a function that can be delegated. While it may be true that, in general, national leaders manipulate and intermediators monitor, this is not invariably the case. The most notable exception arises in the case of teams. The role of a team leader is analagous to that of an intermediator in trade, in that the team leader intermediates between members in the co-ordination of their efforts and in the sharing of the product. He is, though, an intermediator in a multilateral rather than a bilateral sense. Because teamwork brings the leader into a more intimate relation with his followers than does trade, team members are more prone to develop emotional dependence. Thus the team leader can draw upon the moral rhetoric he has learned from the national leader to manipulate team members for his own ends.

While intermediators may have a reputation for integrity they certainly do not have the same reputation for altruism that a national leader may enjoy. In a market economy intermediators will exploit whatever market power their reputation affords. This is reflected in the model presented in Section 8.2. An intermediator negotiates a margin between buying price and selling price which maximizes his share of the gains from trade. One of his key decisions is whether to manipulate his customers or to monitor them. The model examines how far this decision resembles the decisions of national leaders discussed earlier on.

Section 8.3 considers the case where individual intermediators do not have a personal reputation, but rely only on the reputation of intermediators as a group. If not all intermediators are fully committed, then some uncommitted intermediators may attempt to free-ride (see Chapter 10) on the reputation established by the committed ones. If their behaviour undermines confidence in intermediation as a whole then it can have a devastating effect on overall performance. The role of the national leader in maintaining commitment amongst the intermediators is discussed in some detail.

The analysis is extended to team leadership in Section 8.4. The influences of competition and contestability on the rewards to intermediation are discussed in Section 8.5. Section 8.6 summarizes the main results.

8.2. The Intermediator as a Reputable but Selfish Natural Monopolist

Suppose that certain individuals can build a personal reputation for themselves. In the light of earlier remarks about bounded rationality, and the consequent use of oversimplified models, the credibility of this reputation has to be based on a simple and straightforward mechanism. It is therefore assumed that these people are trusted, first and foremost because they are believed to be absolutely committed to the leader's ethic. These followers are somehow able to signal to other followers that they are highly sensitive individuals who would never cheat. While other followers may exhibit a uniform distribution of sensitivity, as assumed earlier, these reputable followers have a sensitivity which is an order of magnitude different.

It is further assumed that the other followers are correct in their beliefs about these select individuals, and that their reputation is fully justified. Although the basic reputation mechanism is personal commitment, this mechanism may, of course, be reinforced by the high profile these individuals are able to assume because of the trust other people place in them. Given their reputation, therefore, they would have a strong incentive to behave honestly even if they lost their commitment, because their misdemeanour would quickly become public knowledge and would thereby undermine their economic status within the group (see Chapter 9).

It was shown in Chapters 4 and 5 that trade can collapse entirely due to lack of trust between transactors. In particular, sensitive but pessimistic traders will refuse to participate in trade, and this will drive up the incidence of cheating amongst those who do participate to a catastrophic level. The existence of reputable individuals resolves this problem. Their reputation means that everyone is happy to trade with them. If they are unhappy to trade with others, then they can demand hostages or resort to moral manipulation.

Within a decentralized economy intermediators often appear as a distinctive social class. To put it crudely, they are a class of middlemen who constitute a 'middle class'. In the models presented below, the leader personifies the 'upper class' and the ordinary followers the 'lower class'. In practice, the social manifestations of these class distinctions are, of course, more prominent in some societies, such as those with low social mobility, than others.

The economic efficiency of a group may depend crucially on the quality of interclass relations. Granted that middlemen may rely on a reputation for integrity, they do not usually enjoy a reputation for altruism as well. This is reflected in the following models, which impute altruism to the leader but selfishness to the middlemen.

Middlemen frequently possess an element of local monopoly power. Because a middleman constitutes an indivisible human resource, he may enjoy a natural monopoly. In a spatially extended economy middlemen may disperse so that each enjoys some monopoly power within his market area and faces competition only at the boundary with other people's market areas.

It is assumed that the total gain from trade, $a>0$ is divided between the middleman and the customer so that the middleman receives $a_1 > 0$ and the customer $a_2 > 0$:

$$a = a_1 + a_2. \tag{8.1}$$

This division is determined by relative bargaining skill and competitive conditions. It is treated as exogenous at this stage, but endogenous later on. Recall that in direct trade each partner sacrifices a good he values at $b > 0$ in return for a good he values at $a + b$. With intermediation each partner sacrifies, as before, a good of value b, but receives in return from the middleman a good of value only $a_2 + b$ (although the good he has supplied to the middlemen is, in fact, of value $a + b$ to the other trader).

The tastes of the middleman, it is assumed, are such that he values each good as a recipient, rather than as a supplier, would do. This is consistent with the view that he is a direct consumer of the goods in which he trades. The middleman physically takes a cut of proportion $a_1/(a + b)$, which reduces what is passed on to a value of only $a_2 + b$ so far as the recipient is concerned. The situation is illustrated schematically in Table 8.1.

TABLE 8.1. *The process of intermediation*

Partner 1		Intermediator		Partner 2	
Sale	Good 1	Purchase	Sale	Good 1	Purchase
$-b$	$\rightarrow$	$a_1 + b$	$-b$	$\rightarrow$	$a_2 + b$
Purchase	Good 2	Sale	Purchase	Good 2	Sale
$a_2 + b$	$\leftarrow$	$-b$	$a_1 + b$	$\leftarrow$	$-b$
a_2		a_1	a_1		a_2

TABLE 8.2. *Information sets for customer and middleman under trade*

A. Customer

Strategy	Reward
Honesty	a_2
Cheating	$a_2 + b - g_2$

B. Middleman

Rewards	Customer's strategy	
	Honesty	Cheating
To middleman	a_1	$-b$
To customer	a_2	$a_2 + b - g_2$
Perceived probability	$1 - p_2$	p_2

The consequences of cheating, as perceived by middleman and customer, are shown in Table 8.2. The subscript 1 identifies the middleman and the subscript 2 the customer. When a customer cheats he conserves the payment which he values at $b > 0$ and so his gain from trade rises to $a_2 + b$. Matters are more complicated for the middleman because he is dealing with two customers in linked transactions. It simplifies the discussion to suppose that when customers cheat they both do so at the same time. It is then clear that the middleman loses b on each transaction compared to the no-trade situation. It is similarly clear that when both middleman and customer cheat the reward to each is zero.

Under manipulation the middleman must set the customer's guilt g_2 at no less than the customer's material incentive to cheat,

$$g_2 \geqslant b. \tag{8.2}$$

With a uniform distribution of sensitivity to guilt, setting the intensity of manipulation in the region of diminishing returns gives a customer crime rate

$$q_2 = b/\theta. \tag{8.3}$$

The middleman's expected reward is

$$v_1 = a_1 - (a_1 + b)q_2 \tag{8.4}$$

and he maximizes

$$u_1 = v_1 - c_1, \tag{8.5}$$

where c_1 represents the usual manipulation costs. Substituting (8.3) and (8.4) into (8.5) and solving the first-order condition gives the optimal strategy

$$\theta_1^e = [(a_1 + b)b/c_{1v}]^{\frac{1}{2}} \tag{8.6.1}$$

$$q_2^e = [bc_{1v}/(a_1 + \mathrm{b})]^{\frac{1}{2}} \tag{8.6.2}$$

$$v_1^e = a_1 - [(a_1 + b)\, bc_{1v}]^{\frac{1}{2}} \tag{8.6.3}$$

$$u_1^e = a_1 - 2[(a_1 + b)bc_{1v}]^{\frac{1}{2}} - c_{1f}, \tag{8.6.4}$$

where c_{1v} represents the middleman's marginal manipulation cost and c_{1f} his fixed cost.

These results have a familiar appearance. They show that the introduction of intermediation does not lead to any radical reconsideration of optimal manipulation strategy. Note how the intensity of manipulation is an increasing function of the middleman's share of the gains from trade, a_1. Conversely, the customer crime rate is a decreasing function of a_1. This leads to the somewhat counter-intuitive result that customers are more honest the less they benefit from trade. The explanation is that as their share of the gains falls, customers do not have any greater material incentive to cheat because their gains from cheating fall by the same amount as do their gains from honest trade. But the middlemen have a greater incentive to manipulate because the marginal return to a reduction in the crime rate increases with their share of the gains from trade. The lesson would seem to be that, in a cross-section of independent economic groups, those customers on whom the most intense manipulation is being targeted are probably those who are most being taken advantage of. This leads to an important qualification to this result—namely that when customers believe that the distribution of gains is unjust, they may experience a strong emotional inducement to rebel. Issues of this kind are considered in Chapter 11.

The alternative monitoring strategy is very straightforward. By requiring that customers made advance payment, the middleman's supplies—which the customers value at $a_2 + b$—can be held hostage against a payment which the customers value at $b < a_2 + b$. Given that the costs of monitoring are independent of the size of the hostage, as assumed in earlier chapters, the use of the middleman's supplies as a natural hostage is an optimal monitoring strategy. It completely eliminates cheating at a fixed cost c'_{1f}.

Given a fixed cost of manipulation c_{1f}, it follows that manipulation is preferred to monitoring only if

$$c_{1f} \leqslant c'_{1f} - 2[(a + b)bc_{1v}]^{\frac{1}{2}} \tag{8.7}$$

Practical considerations suggest that this condition is not usually satisfied and so monitoring will be preferred. It is normally easy for a middleman to detect cheating by customers and easy to implement natural hostages. The main exception is where the customers are remote from the middleman and require their supplies at short notice. In this case the demand for prepayment can cause costly delay. If the customers are regular clients, however, then the problem may be solved by their maintaining a credit balance with the middleman which can be drawn down as and when required. In other cases, however, manipulation may be used as a last resort.

8.3. Middlemen Who May Cheat

A society which has evolved a middle class with a reputation for integrity can face a serious crisis if that middle class then begins to cheat. Loss of reputation by a middle class can be catastrophic for an economy. This section examines the case where middlemen have a collective reputation which falls short of absolute integrity. Instead of each middleman being recognized as personally committed, he is now simply recognized as a member of a specialized functional group. Customers move between different sectors of the economy, and so trade with different middlemen over time. They pool information on middlemen's behaviour.

The middlemen take their cue from the overall leader, and it is his moral manipulation of them that determines the amount of cheating they do. It turns out that under these circumstances the middlemen's choice of strategy is crucial.

If the middlemen rely on manipulation then they will pursue the same manipulation strategy independently of their beliefs about customers and independently of their own sensitivity too. It is only the middleman's decision to cheat that depends on his own sensitivity. This decision is independent of how much customer manipulation he considers worthwhile. This independence results in a relatively stable situation.

If the middlemen rely on monitoring instead, though, then their own incentive to cheat depends critically on their beliefs about the followers. The way the follower's behave depends in turn on what they believe about the middlemen. There is, therefore, in recurrent encounters a mutual interdependency analogous to Cournot reactions. The leader regulates the middlemen's reaction function through moral manipulation. Leader manipulation must attain a critical level if a catastrophe, caused by the decline in middleman reputation, and the knock-on effect on followers, is to be averted.

Consider first the stable, and relatively straightforward case, in which middlemen employ manipulation. The rewards perceived by customers and middlemen respectively are shown in Table 8.3.

TABLE 8.3. *Information sets for customer and middleman: manipulation by middlemen who may cheat*

Customer's strategy	Middleman's strategy	
	Honesty	Cheating
Honesty	a_2	$-b$
Cheating	$a_2 + b - g_2$	$-g_2$
Perceived probability	$1 - p_2$	p_2
Middleman's strategy	**Customer's strategy**	
	Honesty	Cheating
Honesty	a_1	$-b$
Cheating	$a_1 + b - g_1$	$-g_1$
Perceived probability	$1 - p_1$	p_1

It is readily deduced from the table that customers will be honest if, as before, (8.2) is satisfied, whilst middlemen will be honest if

$$g_l \geqslant b. \tag{8.8}$$

The customers' incentive to cheat is independent of how middlemen are expected to behave, whilst the middlemen's incentive to cheat is independent of how customers are expected to behave. What is more, a dishonest middleman's incentive to manipulate a customer is exactly the same as an honest middleman's. Indeed, the optimal strategy for any middleman is given by (8.6) exactly as before. Note that because all middlemen pursue the same strategy, a customer cannot screen a middleman for integrity by observing the manipulation strategy he employs.

A narrowly materialistic altruistic leader, who knows how customers will respond to middlemen's manipulation strategies, will measure performance by

$$v_0 = a - a_1 q_2^e - a_2 q_1. \tag{8.9}$$

The leader maximizes utility

$$u_0 = v_0 - c_0 - c_1. \tag{8.10}$$

where c_0 represents his own manipulation costs, and c_1 the middlemen's manipulation costs, all expressed on a per capita basis with respect to the number of customers. With a uniform distribution of sensitivity among middlemen, their crime rate is

$$q_1 = \begin{cases} 1 & \theta_0 < b \\ b/\theta_0 & \theta_0 \geqslant b . \end{cases} \tag{8.11}$$

Taking $\theta_0 \geqslant b$ to obtain an interior maximum of (8.10), the use of (8.11) gives the first-order condition

$$a_2 b/\theta_0^2 - c_{0v} = 0, \tag{8.12}$$

whence

$$\theta_0^e = (a_2 b/c_{0v})^{\frac{1}{2}} \tag{8.13.1}$$

$$q_1^e = (bc_{0v}/a_2)^{\frac{1}{2}} \tag{8.13.2}$$

$$v_0^e = a - a_1[bc_{1v}/(a_1 + b)]^{\frac{1}{2}} - (a_2 bc_{0v})^{\frac{1}{2}} \tag{8.13.3}$$

$$u_0^e = a - (2a_1 + b)[bc_{1v}/(a_1 + b)]^{\frac{1}{2}} - 2(a_2 bc_{0v})^{\frac{1}{2}} - c_{0f} - c_{1f}, \tag{8.13.4}$$

where c_{0v} is the leader's marginal cost of manipulating a middleman. Since there is no relevant learning about the other party these results are inherently static. In other words, the one-off and recurrent solutions are identical. The main point worthy of note (which must be derived by further analysis) is that if the leader were able to control the middlemen's own manipulation strategy directly, he would set it lower because he is not bothered, as the selfish middlemen are, by the redistribution caused by customer cheating. He would leave his own manipulation of the middlemen unchanged, however, because middleman dishonesty damages the leader's reward by a factor that is independent of customer cheating.

Under monitoring the middlemen require prepayment, and so hold their own supplies as natural hostages. As a result, the reward structures change to those in Table 8.4. The customers will be honest only if middleman cheating is not expected to exceed a critical level:

$$p_2 \leqslant p_2^* = a_2/(a_2 + b). \tag{8.14}$$

This critical value is lower the smaller is the customer's share of the gains from trade. The concept of customer cheating is, in fact, somewhat misleading in this context because customers know that their payment is hostage, so that 'cheating' effectively represents a decision not to participate.

TABLE 8.4. *Information sets for customer and middleman: monitoring by middlemen who may cheat*

Follower's strategy	Middleman's strategy	
	Honesty	Cheating
Honesty	a_2	$-b$
Cheating	0	0
Perceived probability	$1 - p_2$	p_2
Middleman's strategy	**Customer's strategy**	
	Honesty	Cheating
Honesty	a_1	0
Cheating	$a_1 + b - g_1$	$-g_1$
Perceived probability	$1 - p_1$	p_1

Since the middlemen are leader-like rather than follower-like in their behaviour, it is assumed each middleman knows the condition (8.14) and also knows that initial customer probability beliefs are uniformly distributed across the unit interval. Each middleman can therefore correctly predict the first-period incidence of customer cheating:

$$p_{11} = q_{21} = p_2^*, \qquad (8.15)$$

where p_{11} is the middleman's perceived probability of customer cheating in period 1 and q_{21} is the actual customer crime rate in period 1.

Subsequently customers have accurate information on the previous period's cheating across the middleman group as a whole, and the middlemen know this too. The inequality (8.14) then implies

$$p_{1t} = q_{2t} = \begin{cases} 0 & q_{1t-1} \leqslant p_2^* \\ 1 & q_{1t-1} > p_2^* \end{cases} \qquad (t = 2, 3, \ldots) \qquad (8.16)$$

Table 8.4 also indicates that a middleman will be honest only if his guilt reaches a probability-dependent threshold level related to his material incentive to cheat

$$g_1 \geqslant (1 - p_1)b. \qquad (8.17)$$

The appearance of the term p_1 in (8.17) reflects an assumption that in each period middlemen must determine their intentions before they know whether customers have placed a hostage or not. When few hostages are expected the middlemen incline to honesty because the material gain to cheating is low while the emotional penalty from the intention to cheat those who do not actually place hostages remains high. Conversely, when many hostages are placed the middlemen are inclined to cheat, because the material gain is large, relative to the emotional penalty involved. Assuming a uniform distribution of sensitivity amongst the middlemen, the application of (8.15) to (8.17) gives the first-period crime rate

$$q_{11} = \begin{cases} 1 & \theta_0 < (1 - p_2^*)b \\ (1-p_2^*)b/\theta_0 & \theta_0 \geqslant (1 - p_2^*)b. \end{cases} \qquad (8.18)$$

Subsequently middlemen have information on the actual follower crime rate for the preceding period, whence

$$q_{1t} = \begin{cases} 1 & \theta_0 < (1 - q_{2t-1})b \\ (1-q_{2t-1})b/\theta_0 & \theta_0 \geqslant (1 - q_{2t-1})b. \end{cases} \qquad (t = 2, 3, \ldots) \ (8.19)$$

With one-off encounters the overall leader seeks to maximize

$$u'_{01} = v'_{01} - c_0 - c'_1, \tag{8.20}$$

where

$$v'_{01} = a - a_1 q_{21} - a_2 q_{11}, \tag{8.21}$$

c_0 is the leader's manipulation cost, and c'_1 the middleman's monitoring cost. Substituting (8.15), (8.18) and (8.21) into (8.20) and seeking an interior maximum with respect to θ_0 gives

$$\theta^e_{01} = b[a_2/(a_2 + b)c_{0v}]^{\frac{1}{2}} \tag{8.22.1}$$

$$q^e_{11} = b[c_{0v}/a_2\,(a_2 + b)]^{\frac{1}{2}} \tag{8.22.2}$$

$$v'^e_{01} = a - [a_1 a_2/(a_2 + \mathrm{b})] - b[a_2 c_{0v}/(a_2 + b)]^{\frac{1}{2}} \tag{8.22.3}$$

$$u'^e_{01} = a - [a_1 a_2/(a_2 + b)] - b[a_2\, c_{0v}/(a_2 + b)]^{\frac{1}{2}} - c_{0f} - c'_{1f}, \tag{8.22.4}$$

where c_{0v} is the leader's marginal cost of manipulation, c_{0f} his fixed cost, and c'_{1f} the middleman's fixed cost of monitoring (on a per capita basis as before).

Comparing (8.22) with (8.13) shows that the intensity of leader manipulation is lower with monitoring and the crime rates are also lower. The customer crime rate is lower because all customer crime is eliminated, and the middleman crime rate is lower because an honest middleman no longer has much to fear from a customer who cheats.

With recurrent encounters the leader's role becomes particularly important, because it is necessary to engineer an equilibrium within a potentially unstable system. Combining (8.16) and (8.19) shows that over time there are two interleaved sequences. One begins with the middlemen's initial decisions and the customer's responses, and the other with the customer's initial decisions and the middlemen's responses. These interleaved sequences are independent of one another, and so for equilibrium both must converge to the same state. Fortunately, the logic of the sequences is the same despite the fact that they start differently, so that one converges only if the other does and both converge to the same point.

If $\theta_0 > 0$ there can be no equilibrium where customers cheat, because if customers cheat middlemen will be honest (as they feel guilty, and there is no material gain to cheating when a hostage has not been placed). If all middlemen are honest then all customers will be honest (since it is expedient to be so), and so customer cheating implies customer honesty two periods later. When customers are monitored rather than manipulated they all behave alike, and hence if customers do not cheat they are all honest instead. They will continue to be honest provided middlemen do not cheat too much. (There is a potential information problem in that if all customers cheat, middlemen receive no hostages and cannot therefore demonstrate their honesty, but this is ignored.)

Middleman cheating in response to honest customers generates a crime rate (from (8.19) with $q_{2t-1} = 1$)

$$q_1 = b/\theta_0 \tag{8.23}$$

and this does not exceed the critical value p_2^* if

$$\theta_0 \geqslant \theta_0^* = (a_2 + b)(b/a_2) \tag{8.24}$$

where

$$\theta_0^* > b. \tag{8.25}$$

In the steady state where customers are honest the leader's reward is simply

$$v_0' = a - a_2 q_1 \tag{8.26}$$

and the leader's optimal strategy if the constraint (8.25) is not binding is to set

$$\theta_0^e = (a_2 b/c_{0v})^{\frac{1}{2}}, \tag{8.27.1}$$

so that

$$q_0^e = (bc_{0v}/a_2)^{\frac{1}{2}} \tag{8.27.2}$$

$$v_0'^e = a - (a_2 bc_{0v})^{\frac{1}{2}} \tag{8.27.3}$$

$$u_0'^e = a - c_{0f} - c_{1f}' - 2(a_2 bc_{0v})^{\frac{1}{2}}, \tag{8.27.4}$$

where c_{1f}' is the middleman's fixed monitoring cost as before. Equations (8.27.1) and (8.27.2) indicate that the leader's strategy is exactly the same as in the manipulation case [equations (8.13)].

If however,

$$c_{0v} > (a_2 + b)^2(b/a_2), \tag{8.28}$$

then the constraint (8.25) is binding, and

$$\theta_0^e = \theta_0^* \tag{8.29.1}$$

$$q_1^e = p^* \tag{8.29.2}$$

$$v_0'^e = a - [a_2^2(a_2 + b)] \tag{8.29.3}$$

$$u_0'^e = a - c_{0f} - c_{1f}' - [(a_2^2(a_2 + b)] - [(a_2 + b)bc_{0v}/a_2]. \tag{8.29.4}$$

The derivation of the solution is illustrated in Fig. 8.1. The customer's response function is given by the discontinuous schedule $OR_2R_2'Z$. With $\theta_0=b$ the middlemen's response function is the reverse diagonal R_1R_1'. There is no equilibrium with this minimal intensity of manipulation.

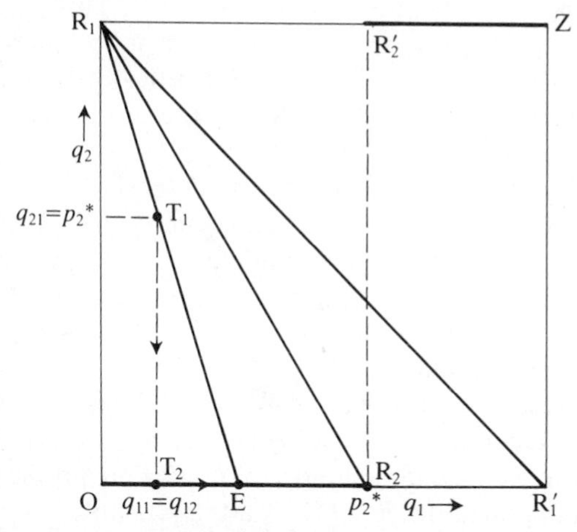

FIG. 8.1 Equilibrium and stability with a leader who manipulates middlemen who monitor customers

The leader cannot influence the customer's response function but he can rotate the middlemen's response function clockwise about R_1 by increasing the intensity of manipulation. The line R_1R_2 corresponds to the critical level of manipulation θ_0^* which sustains middleman cheating at p_2^*. This gives an equilibrium at the intersection R_2 of R_1R_2. An interior equilibrium is achieved with a still higher intensity of manipulation. The middlemen's response function rotates from R_1R_2 to R_1E and gives an equilibrium at E where the middlemen's crime rate is significantly less than the critical level p_2^*.

The dynamics of adjustment to E are illustrated by the trajectory T_1T_2E in the figure. In the first period the incidence of customer cheating is equal to the critical value of the middlemen's crime rate $q_{21} = p_2^*$ while the middlemen, correctly anticipating this, generate a crime rate q_{11}. The first period outcome is thus represented by T_1. Since the middlemen's crime rate is below the critical value, the customers respond in the second period by total honesty, $q_{22} = 0$. The middlemen, having had their expectations confirmed, continue as before, $q_{12} = q_{11}$, giving the second period outcome T_2. In the third period the customers see no reason to change their strategy, since the middlemen did not change theirs the previous period, so that $q_{23} = 0$. But the middlemen now change their strategy in response to the elimination of customer cheating, increasing their crime rate to q_{13} to take advantage of honest customers. This pair of values corresponds to the equilibrium E. It is an equilibrium because the middleman crime rate is still below the critical value, so that customers remain honest, and middlemen therefore have no further reason to change their own strategy.

It is assumed that the leader is able to determine the institutional framework within which trade occurs and hence to dictate the choice between monitoring and manipulation. If middlemen were to have a free hand in this, then different decisions might be taken by different middlemen according to their sensitivity and the intensity of leader manipulation. It avoids this complication if the leader's discretion is assumed throughout. With middlemen acting as manipulators the maximum leader utility is given by (8.13.4). Assuming, for simplicity, an interior solution to monitoring problems, where (8.25) is not binding, then the relevant leader utility is (8.27.4). A direct comparison indicates that the leader will prescribe manipulation rather than monitoring if

$$c_{1f} < c'_{1f} - (2a_1 + b)\,[bc_{1v}/(a_1 + b)]^{\frac{1}{2}} \qquad (8.30)$$

Manipulation is likely to be preferred to monitoring, therefore, the lower is the material incentive to cheat, b, and the smaller the middleman's gain from trade a_1. When middlemen obtain a large gain from trade they tend to overinvest in manipulation so far as the leader is concerned.

8.4. Team Leadership

One of the roles of an employer is to intermediate between the team members he employs. His role is closely analogous to that of a middleman though there are, of course, differences because of the technological complementarities involved. The distribution of gains from teamwork is determined by the wage that is paid.

Because the term 'employer' has a number of other connotations besides the intermediation of team production, we use the term 'team leader' instead. Under manipulation, the team leader assures each member of a fixed wage, $w > 0$. Following Chapter 7, team output when everyone is dedicated is y and when some people slack is $y - \Delta y > 0$. The cost of effort to a member is $e > 0$.

For the sake of simplicity, only the case of perfectly correlated probabilities of slacking is considered. Because each member is now assured of a fixed wage, it is the leader, as residual claimant, and not the team member that is concerned about the incidence of slacking. Since the team leader is assumed to have true knowledge of the followers, the leader will perceive perfect correlation only if it is true. In general, slacking decisions will be perfectly correlated under manipulation only if all members have the same sensitivity. It is therefore assumed that in any given team all members have the same sensitivity, despite the fact that membership is generated by sampling from a population in which sensitivity is uniformly distributed. The leader knows that all members will have the same sensitivity, but he does not know what that sensitivity will be. He knows only that it is uniformly distributed across the population as a whole.

TABLE 8.5. *Information sets for team member and team leader*

A. Team member

Strategy	Reward
Honesty	$w - e$
Cheating	$w - g_2$

B. Middleman

Rewards	Member's strategy	
	All honest	Some cheating
To middleman	$y - w$	$y - \Delta y - w$
To member	$w - e$	$w - g_2$
Perceived probability	$1 - p_1$	p_1

Table 8.5 illustrates the distribution of rewards when team leaders have absolute integrity. A member is dedicated if

$$g_2 \geqslant e, \tag{8.31}$$

which means, for any intensity of manipulation $\theta_1 \geqslant e$, that the incidence of slacking is

$$q_2 = e/\theta_1. \tag{8.32}$$

The team leader maximizes

$$u_1 = v_1 - c_1, \tag{8.33}$$

where

$$v_1 = y - w - \Delta y q_2 \tag{8.34}$$

and c_1 represents the usual manipulation costs.

The customary first-order condition implies

$$\theta_1^e = (e\Delta y / c_{1v})^{\frac{1}{2}} \tag{8.35.1}$$

$$q_1^e = (ec_{1v}/\Delta y)^{\frac{1}{2}} \tag{8.35.2}$$

$$v_2^e = y - w - (e\Delta y c_{1v})^{\frac{1}{2}} \tag{8.35.3}$$

$$u_1^e = y - w - 2(e\Delta y c_{1v})^{\frac{1}{2}} - c_{1f}, \tag{8.35.4}$$

where c_{1f}, c_{1v} are the fixed and marginal costs of manipulation. Once again there are no learning effects and so the results are the same in both one-off and recurrent situations.

The results (8.35) are strikingly similar to the results (8.6) obtained in the case of trade in Section 8.2. This is because the introduction of the wage system establishes a set of separable bilateral contracts between the team leader and each member of the team. The converse of this is that the results are very different to the results (7.5) for the altruistic leader of team production in Section 7.3. Assuming $e < \Delta y$, the intensity of manipulation is much greater, and so too is the incidence of slacking. This is because of the effort disincentive which is implicit in replacing a share of joint output with a fixed wage. The poorer performance is reflected in the fact that the team leader obtains smaller reward, and lower utility, than does the altruistic leader, even when the wage is set to zero.

Once again a monitoring system will serve to eliminate cheating altogether. It is assumed that the leader relates bonuses and fines to individual rather than team performance so that members continue to be insured against the consequences of slacking by others. If the wage is large enough then the payment of wages in arrears can be used to create a natural hostage. Thus if the wage exceeds the cost of effort, $w > e$, and no one is paid if they slack, everyone will be dedicated. As in Chapter 7, such behaviour validates the assumption that slacking decisions are perfectly correlated. If c'_{1f} is the fixed cost of monitoring, and the marginal cost is zero as before, then the condition for manipulation to be preferred to monitoring is

$$c_{1f} < c'_{1f} - 2(e\Delta y c_{1v})^{\frac{1}{2}}. \tag{8.36}$$

Thus manipulation is most likely to be preferred when the cost of effort, the marginal productivity of effort and the fixed and marginal costs of manipulation are all low.

Team leadership can also be analysed in the case where leaders are liable to cheat, but no new insights emerge. Once again, there is a critical level of integrity that is required to sustain a desirable equilibrium. Team leaders, like their middlemen counterparts, need to maintain a good reputation with their followers if they are to carry out their role effectively.

8.5. Competition

When an intermediator has an uncontested monopoly he has a strong incentive to squeeze the followers by maximizing his own share of the gains. When the intermediator has a reputation for perfect integrity then altering the distribution of rewards does not normally affect the follower's incentive to cheat. Under manipulation, squeezing the follower reduces the reward from cheating as much as it does the reward from honesty, so the follower has no incentive to modify his behaviour. Under monitoring, however, the value of the natural hostage may be reduced as a result, and so a special hostage may be required instead. Subject to this qualification, the assumption that the marginal cost of monitoring is independent of the size of hostage means that hostage value can be set as high as is necessary in order to eliminate the incentive to cheat. Thus, whatever the distribution of rewards, a monitored follower will not cheat.

The implication is that an intermediator with an uncontested monopoly will appropriate all the gains from trade for himself. Thus, when acting as a middleman he will maximize the spread between buying price and selling price by setting

$$a_1 = a; \quad a_2 = 0, \tag{8.37}$$

while when acting as team leader he will set the wage at the subsistence limit w:

$$w = \bar{w}. \tag{8.38}$$

By contrast, when the monopoly is contestable the intermediator must defend his position again potential entrants. If the overall leader licenses the right to intermediate then competitive bidding amongst equally skilful potential intermediators will cause all the rent to accrue to the leader. Followers will be squeezed just as before, intermediator utility will be zero, and surplus gains (net of monitoring or manipulation costs) will go into the leader's pocket.

The more interesting case is where rival intermediators compete to serve the followers. Although the dynamics of competition in this case can be quite complicated, the overall tendency is clearly to eliminate intermediator utility through a redistribution of rewards to the followers. The distribution of rewards is obtained simply by equating the appropriate equilibrium utility to zero and solving for the value of the distribution parameter.

In the case of a middleman, setting $u_1^e = 0$ in (8.6.4) gives the margin generated by competition between potential manipulators. Solving the relevant quadratic and taking the positive root (as required for economic relevance) gives

$$a_1^e = c_{1f} + 2bc_{1v}\,(1 + b + c_{1f} + bc_{1v}). \tag{8.39}$$

As expected, the competitive value of the leader's margin is an increasing function of the leader's costs of manipulation c_{1f}, c_{1v} and the followers' material incentive to cheat.

In the case of teamwork the competitive wage can be derived more easily. This is because the level of wage does not affect the team leader's incentive to manipulate in the same way that the middleman's margin affects his incentive to manipulate. Rearranging equation (8.35.4) shows that the competitive wage is

$$w^e = y - c_{1f} - 2(e\Delta yc_{1v})^{\frac{1}{2}}. \tag{8.40}$$

The wage is thus an increasing function of potential team productivity y and a decreasing function of manipulation costs c_{1f}, c_{1v}, the cost of effort e, and the marginal productivity of effort Δy.

8.6. Summary

The manipulation of customers by middlemen, and of team members by team leaders, strongly resembles the manipulation of a national economy by a solitary leader. There are, however, a number of complicating factors which need to be noted. The main difference from national leadership is that because intermediators operate in a market environment and are motivated by selfish ends, their manipulation strategy is adapted to the competitive environment. The intensity of manipulation is higher, and the crime rate consequently lower, the larger the share of gains that they appropriate. Although the material incentives for customers and team members to be honest are lowered when most of the gains are syphoned off, the intermediator's intensified manipulation increases emotional penalties for cheating to a level which more than compensates for this. This shows that selfish manipulators can redistribute income in their favour, not only by clever bargaining but by inculcating greater honesty in those they are taking advantage of.

By and large, middlemen will prefer monitoring to manipulation because it is easy for them to demand hostages (by insisting on pre-payment of sales and post-payment of purchases) and so eliminate all exposure to risk from customer cheating. Where team leadership is concerned the case for monitoring is not so clear-cut because the monitoring of individual effort is often costly and the monitoring of team output requires team-based incentives to be used (as indicated in Chapter 7).

The middleman's choice between monitoring and manipulation is an important issue for the national leader when middlemen are potentially dishonest. Consider the case of manipulation first. The optimal manipulation strategy of a middleman is independent of his own moral commitment. Customers stand to gain the same amount from cheating him whether he is honest or not, and he stands to gain the same amount from cheating them as well. Over time they will not respond to his behaviour and he will not respond to theirs. Thus the pattern of behaviour in one-off and recurrent trades is the same. The leader cannot influence customer behaviour because he cannot influence the intensity of middleman manipulation. He can only influence the incidence of middleman cheating. The middleman's intensity of manipulation varies directly with the customer's share of the gains from trade and the customer's material incentive to cheat. If the national leader could, in some way, directly control the middlemen's manipulation strategy he would set their intensity of manipulation lower because he would be less concerned than they are with reducing the redistribution of income to customers by deterring customer cheating.

When middlemen employ monitoring, on the other hand, the customers become more vulnerable and the middlemen more inclined to cheat. If middlemen's cheating exceeds a critical level then customers will refuse to trade. On the other hand, the more honest the customers are, the greater is the incentive for the middlemen to cheat. This is because dishonest customers place no hostages, and so provide no material gains from cheating, although middlemen continue to experience guilt from knowing that they would cheat them if they did. Thus as customer honesty increases the relative weight carried by material gains in middlemen's calculations increases, and the incidence of middlemen's cheating rises as a result. The complex interaction between middlemen's and customers' expectation of each other's honesty generates oscillations in the crime rate in recurrent trades. Astute national leadership is required to ensure that in the long run middlemen's honesty is maintained above the critical level. The optimal strategy relates the intensity of manipulation directly to the middlemen's incentive to cheat and to their share of the gains from trade, and inversely to the leader's cost of manipulation.

9

Small is Cosy: Intimate Relations in Small Groups

9.1. Introduction

The moral manipulation of people's emotions is generally much easier in small, stable, and compact groups (Chell 1985; Schein 1985). It is also less necessary, because regular contact between the same small number of people encourages integrity purely out of self-interest (Fudenberg and Kreps 1987; Radner 1985; Trivers 1971). On both counts, therefore, small groups are likely to perform better than large ones. Large groups, of course, permit a more sophisticated division of labour and promote the exploitation of technological economies of scale. But this market-oriented technology-driven view ignores transactions costs, and the way that transactions costs are governed by the characteristics of the social group.

There are four main reasons why transaction costs will be lower in a small stable compact group.

1. Personal reputation is easy to acquire. Someone who cheats will be expected to cheat again, and so people will avoid dealing with him in the future if they can. Cheats are therefore punished by being deprived of future opportunities, and potential cheats who recognize this may be honest out of pure expediency.

2. Positive feelings of satisfaction (as described in Chapter 6) are easier to generate. Emotional intensity is raised when a small number of people exist side by side for a prolonged period of time. When personal relationships are good, emotional bonds easily become focused around a common cause, or mission, that the group pursues. This natural process of focused bonding reduces the cost of manipulation. It also makes the group attractive to potential recruits

and so allows the leader to be very selective in who is allowed to join the group (see Chapter 13).

3. The heightened emotional intensity means that anger is easily aroused in small groups. At the same time revenge becomes easier to implement. Those who cheat not only risk loss of reputation, therefore, but they are more likely to be punished directly by their victim too.

4. The leader's costs of manipulation are reduced when people are close together. The quality of information about the followers, which is used to fine-tune manipulation policy, is likely to be better too.

9.2. Personal Reputation in Small Groups

One of the features of a stable group is that learning can occur. In Chapter 4 this learning involved people modifying their opinions about the average behaviour of the group. The group acquired a reputation with itself which was modified over time in the light of the most recent experience.

Reputation was focused on the group rather than on individual members because it was assumed that the group was large. With random (and therefore serially independent) selection of partners, and encounters restricted to unit frequency, the large size of group implies that the chances of meeting any partner again within a given period of time are small. In a small group, however, the chances of meeting again are much higher (assuming that the per capita frequency of encounters is the same as before). Thus it is worthwhile for individuals to record one another's behaviour in order to establish a reputational profile on each fellow-member of the group.

Reputational information, like other information, has public good characteristics. This means that individuals can benefit by sharing their experiences about encounters with other people, provided that the costs of interpersonal communication are not too high. Given that people are systematically collecting information about each other, it may also pay them to eavesdrop on encounters between other people. This eavesdropping has a dual role. First, it accelerates the diffusion of information about behaviour, since the primary sources are no longer just the parties involved but the observers too. Secondly, it not only increases the quantity of information available to any individual, but the quality too. The quality of information is increased because the presence of independent witnesses means that the opportunity for those directly involved in an encounter to misrepresent what occurred is decidedly limited. The existence of multiple sources allows information to be cross-checked.

Opportunities for eavesdropping and gossiping are greatest when the group is spatially compact, so that members reside and work in close proximity to each other. Thus the small, stable, and compact group is particularly advantaged so far as the cultivation of individual reputations is concerned.

The effects of personal reputation are most readily examined within the context of a trading group. It is assumed that participation in trade is voluntary, and the material incentive to default is independent of the partner's strategy. The basic model was outlined in Chapter 5 and the relevant reward structure is reproduced in Table 9.1.

TABLE 9.1. *Follower's perceived rewards for participating in trade under the guilt mechanism*

Strategy	Partner's strategy	
	Honesty	Cheating
Honesty	a	$-b$
Cheating	$a + b + g$	$-g$
Avoidance	0	0

Expectations are again based on past experience. In the first period expectations are purely subjective and follow a uniform distribution across the group (the indivisibility of members becomes more important in a small-group situation, but this problem is ignored). In subsequent periods people are expected to cheat if they cheated in the previous period. If they did not participate in the previous period then their reputation is based on their last recorded behaviour. If they join in later than others, without any previous experience, their reputation is based on the recorded average crime rate of the last contingent to join; this reflects members' naïve belief that those who join later are similar to those who joined immediately before.

With an intensity of manipulation that equals or exceeds the material incentive to cheat, $\theta \geqslant b$, the first-period outcome is that shown in Fig. 9.1. By the end of the period each of the participants has a reputation for either honesty ($p = 0$) or cheating ($p = 1$). The area of the rectangle GBDA′ represents those known for their honesty and the area of the quadrilateral OBDC′ those known as cheats. The non-participants are represented by the area of the quadrilateral A′DC′G′. Their reputation is based on the average crime rate, which is given by the ratio of the area OBDC′ to the total area OGA′DC′.

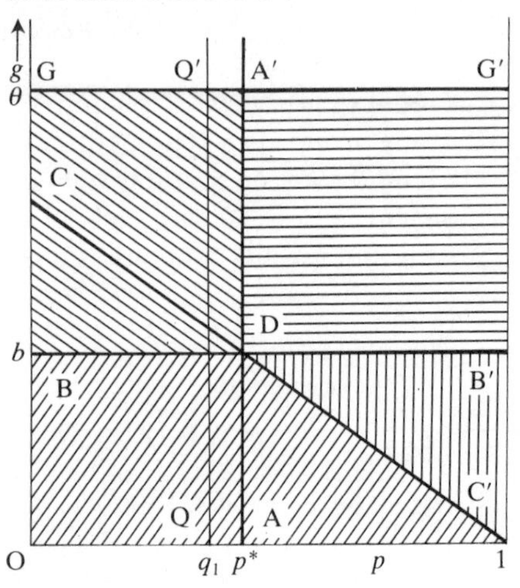

FIG. 9.1 Effects of personal reputation when first-period experience induces everyone to attempt participation in period two

Everyone prefers to avoid a cheat, because neither honesty nor cheating can improve upon avoidance. Since a cheat cannot, therefore, get anyone to trade with him again, he cannot live his bad reputation down. 'Give a dog a bad name and you may as well hang him' describes his situation well. He is excluded from trade in perpetuity.

Suppose to begin with that no one recognizes this effect. Those in the area OBDC′ do not realize until too late that in the second period they are excluded from trade. Honest people can continue to trade with each other without any difficulty, however. So long as the intensity of manipulation is not reduced they will remain honest and their reputation will be kept intact.

Overall behaviour in the second period depends upon how many of the previous non-participants decide to join in. If the first-period crime rate is less than the critical value required to induce honest participation,

$$q_1 \geqslant p^* = b/(a + b), \tag{9.1}$$

then everyone decides to participate. All previous non-participants are more optimistic than they were before. Those who in the first period lay within the triangle DBC′ start cheating whilst those within the rectangle A′DB′G′ start trading honestly. By the third period the new cheats are eliminated by their bad reputation and all the remaining participants are honest. These participants are represented by the area of the rectangle BGG′B′; they constitute a proportion $1 - (b/\theta)$ of the membership. Thus equilibrium is achieved with a zero crime rate.

$$q^* = 0, \tag{9.2.1}$$

and an avoidance rate governed by the intensity of manipulation and the material incentive to cheat:

$$x^* = b/\theta. \tag{9.2.2}$$

If, on the other hand, the first-period crime rate exceeds the critical value given in (9.1) then only some of the previous non-participants will join in the second period. Those who join are people with low sensitivity who were previously very pessimistic but are now a little less pessimistic than before. Fig. 9.2 illustrates the case where a lower intensity of manipulation $\theta^{\dagger} < \theta$ induces a higher crime rate $q_1^{\dagger} > q_1$ and where, in consequence, only members located in the triangle HJK are induced to join in the second period. All of the new participants join only to cheat. In the light of the earlier assumptions, they give all the non-participants the reputation of cheats. Thus the actual cheats are eliminated by the third period by their proven misdeameanour whilst the remaining non-participants find it impossible to join because of guilt by association—the honest will not trade with them and since they share other people's

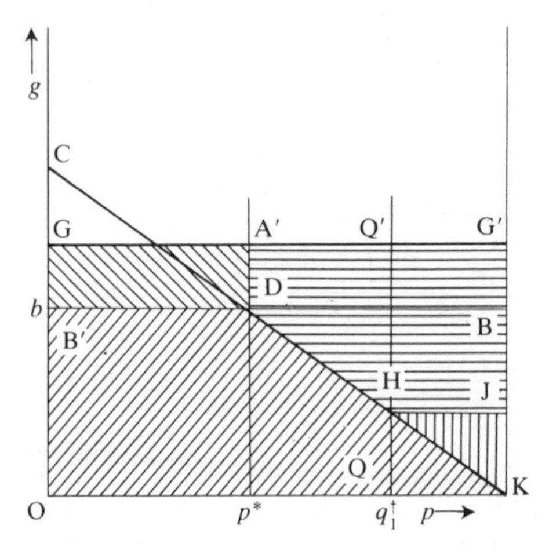

indicates cheating from period one

indicates honesty from period one

indicates cheating from period two, after avoidance in period one

indicates non-participation due to avoidance and later exclusion

FIG. 9.2 Effects of personal reputation when first-period experience induces everyone to attempt participation in period two

estimation of themselves they will not trade with each other either. Participation is therefore restricted to the area of the rectangle GA′DB′. The equilibrium crime rate is again zero

$$q^{\dagger *} = 0, \tag{9.3.1}$$

but the avoidance rate is increased to

$$x^{\dagger *} = 1 - [b/(a + b)][1 - (b/\theta)] > x^*. \tag{9.3.2}$$

The results have a certain affinity with those derived in Chapter 3 where reputation operates at the group level and participation in trade is compulsory. But while in Chapter 3 the avoidance rate was zero and the crime rate was governed by the intensity of manipulation, in the present case the crime rate is zero and the avoidance rate depends on the intensity of manipulation. In fact, the avoidance rate varies with the intensity of manipulation in the same way that the crime rate did before.

From an altruistic leader's point of view the way these reputation effects work out is not particularly desirable. The main effect is to transform cheating into non-participation rather than into honesty. Indeed, if the distributional consequences of cheating are ignored, a given level of cheating with full participation is no better than complete honesty with reduced participation (because q^* and x^* enter symmetrically into the leader's objective (5.5)). If everyone except the cheat participates then the result is the same, whereas when some are excluded from participating by reputation alone the result is actually worse.

9.3. Incentives Created by Belief in Reputation Effects

The resolution of this paradox is effected by recognizing that the efficiency gains from reputation come when people recognize how the reputation mechanism affects themselves. It is when people realize that if they cheat they will be excluded from subsequent trades that their perceived incentive structure changes and potential cheats are encouraged to be honest instead. Each individual realizes that his share of future gains from trade is hostage to his current action. Exclusion becomes a threat instead of a reality, as individuals substitute towards honesty in order to continue to participate in future.

Such substitution will occur even in the absence of manipulation, provided the value of the hostage at the time of decision is sufficiently large. The value of the hostage depends on the gains from trade, the number of subsequent trades the individual hopes to participate in whilst a member of the group, and the time discount factor he applies to future, as compared to present, gains from trade.

In the absence of manipulation an individual facing the incentives described in Table 9.1 is convinced that at some stage he will wish to cheat. But he knows that this cheating will constitute his last transaction within the group. The key issue is whether he should cheat immediately or defer. For in a stable environment, if it pays to defer now then it pays to defer again later, and so on until the individual knows that his last chance to cheat has come. But if no new information about the environment becomes available, he never knows when his last chance will be, and so he will defer indefinitely. Thus, under these conditions, deciding whether to defer is effectively deciding whether to cheat or not.

An individual can defer cheating either by being honest in the current period, or by avoiding trade. Avoidance is obviously inefficient if the discount rate is positive, $r < 0$, because avoidance generates a zero reward in the first period. Honest participation is better so long as the expected crime rate is less than the critical value p^*. In this case, by deferring cheating the individual expects to receive $a - p(a + b)$ immediately from honesty and $(1 - p)(a + b)$ from cheating in the following period, assuming a suitable opportunity arises. If he does not defer cheating he receives $(1 - p)(a + b)$ immediately and nothing thereafter. Given a probability $i(0 \leqslant i \leqslant 1)$ that he will have parted company with the group in the mean time, present value of deferral is

$$[(1 - p)(1 - i)(a + b)/(1 - r)] - b,$$

which is positive if

$$r < [1 + (a/b)](1 - i)(1 - p) - 1. \tag{9.4}$$

Thus a low rate of time discount, r, promotes honesty without manipulation. So too, for obvious reasons, does a large gain from trade, a, a low material incentive to cheat, b, a low expected crime rate, p, and a low probability of parting company from the group, i.

For a sufficiently small r, a leader can guarantee total honest participation within his group. By announcing that everyone will be honest, the leader sets $p = 0$. In a totally stable group ($i = 0$) everyone will be honest in the first period provided that

$$r < a/b. \tag{9.5}$$

The announcement is self-validating and immediately establishes an equilibrium with

$$q^* = 0; \; x^* = 0. \tag{9.6}$$

For an altruistic leader this equilibrium completely dominates all the others so far considered, provided the cost of the announcement is not too high.

An interesting feature of this equilibrium is that members' beliefs about how reputations work are never actually put to the test. The belief in reputation is entirely sufficient to produce the desired effect. This suggests that, in general, an effective leadership strategy may be to announce how the mechanism is supposed to work, even if in fact it does not, or cannot, work in that way. It is, however, a risky strategy in practice, for if proved wrong, the leader's own reputation may well be ruined, and the potential power of future announcements nullified.

9.4. Size, Dispersion, and Communication Cost

Communication within a group can be effected on either a customized or a standardized basis. Under customization the questions asked by the leader are tailored to the follower's specific. responsibilities and the moral rhetoric is adapted to the follower's particular doubts and anxieties. Each follower is allowed to respond in his own idiosyncratic way. With standaardization, by contrast, the same questions and the same moral arguments are addressed to all followers and they are required to respond in a standardized format.

It is fairly obvious that customization is impractical in a large group unless extensive delegation to intermediaries is employed (Arrow 1974). Equally it is clear that the set-up cost of devising a standardized format cannot be spread over many members in a small group, so that the cost of standardization is fairly high. Table 9.2 indicates that, other things being equal, customization will be selected in small groups and standardization in large ones. This will apply whether it is moral manipulation or monitoring that is involved.

TABLE 9.2. *Average costs of alternative communication strategies, by size of group*

Type of group	Type of communication	
	Customized	Standardized
Small	Low	Medium
Large	High	Low

Other things are not always equal, however, because small groups are typically more compact than large ones. Differences in dispersion affect the choice between face-to-face and impersonal communication. This is important because in this particular respect manipulation and monitoring are not equally affected. Dispersion inhibits monitoring less than it does manipulation.

The basic tendency, of course, is for compact groups to rely on face-to-face communication and dispersed groups to rely on impersonal communication. This is because the costs of face-to-face communication are highly sensitive to dispersion whereas the costs of impersonal communication are not. This is most evident in the case of monitoring summarized in Table 9.3. When face-to-face communication is used in dispersed groups it is normally cheapest for the followers to converge on the leader for otherwise the leader's itinerary would absorb too much of his valuable time—a consideration which is exaggerated by a large size of group. Even in a small group, however, the additional cost will normally be considerable compared with the analogous cost incurred by impersonal communication.

TABLE 9.3. *Average cost of alternative communication strategies for monitoring, by dispersion of group*

Type of group	Type of communication	
	Face-to-face	Impersonal
Compact	Low	Medium
Dispersed	High	Medium

It is important to realize that feelings can be aroused more readily by face-to-face contact than by remote communication. Since manipulation depends on the emotions, whereas monitoring does not, this means that where manipulation is concerned, face-to-face contact is often cheaper than impersonal communication even in a large group (see Table 9.4). Impersonal communication is quite ineffectual in conveying moral sentiments, and so large group manipulation must normally rely on face-to-face contact just as in a small group.

TABLE 9.4. *Average cost of alternative communication strategies for manipulation, by dispersion of group*

Type of group	Type of communication	
	Face-to-face	Impersonal
Compact	Low	High
Dispersed	High	Very high

The implications of this discussion are summarized in Table 9.5. Comparing the entries on the diagonal shows that as a small

TABLE 9.5. *Least-cost communication strategies*

Dispersion of group	Size of group	
	Small	Large
Compact	Customized face-to-face	Standardized face-to-face
Dispersed	Customized *either* Impersonal monitoring *or* Face-to-face manipulation	Standardized *either* Impersonal monitoring *or* Face-to-face manipulation

compact group evolves into a larger more dispersed one, monitoring mechanisms become more standardized and impersonal. In a small compact group the leader will typically supervise the members by observing their activities at first hand. He may also receive verbal reports, which he can check out by inspection and cross-examination on the spot. In a large and diversified group, on the other hand, the leader will tend to rely upon written reports. Instead of allowing people to report about the things that matter to them, in the manner they like, a standard format is prescribed to make it easier for the leader to interpret all the information. The standardized format, however, normally increases the amount of time that members must devote to compiling their reports; and it tends to increase the supply of irrelevant information and to suppress the reporting of unusual events. It can also create a bias towards quantitative information which, though trivial, is easy to communicate and corroborate, and away from qualitative judgement which is difficult to convey and to check out except on a customized face-to-face basis.

Where manipulation is concerned, the transition to a large dispersed group has a similar effect on standardization, but much less effect on impersonality. The leader of a small group will manipulate through personal example and through personal dialogues in which he explains his moral standpoint and answers each follower's objections in the course of conversation. In a larger group the role of personal example is diminished, and the moral message is more standardized. To be intelligible to all the leader needs to aim at the 'lowest common denominator', so that his message may degenerate into propaganda of a rather crude kind.

In a large group the moral standpoint will be articulated more at assemblies and rallies than in one-to-one conversations. If dispersion is only modest then the assemblies may all be held at the leader's headquarters to conserve the leader's travelling time, but where it is very dispersed the leader may travel around to regional assemblies where he repeats his message.

The trend to a propaganda is not the only reason why the message may change as manipulation is extended to a larger group. It was noted in Chapter 6 that feelings of satisfaction from mutual honesty can be engineered as well as feelings of guilt. The emotional intensity of small-group relations means that positive feelings of satisfaction may be particularly easy to engineer when the group is small. Thus small-group manipulation may concentrate on the benefits of mutual honesty, while large-group manipulation plays more on the sense of guilt.

Other differences between small-group and large-group manipulation are possible too. Revenge is much easier in small groups than in large groups, for reasons indicated in Section 9.1. While this may have its advantages in deterring cheating, it means that if cheating does occur a feud may quickly develop. The ease with which an offender can be identified, and the speed with which he can be punished, means that revenge may well be taken in the heat of the moment, rather than after a delay in which the emotions have cooled off. Emotional over-reaction or punishment can turn the offender into a victim and so perpetuate the process. Moral manipulation in small groups may therefore have to place greater emphasis on cooling tempers and stopping feuding than is necessary within a larger group.

9.5. The Impact of Size and Dispersion on the Choice of Co-ordination Strategy

So far as the choice between monitoring and manipulation is concerned, the overall implication is that manipulation is most likely to be chosen in small groups. For any given size and dispersion of group, an optimizing leader will compare the least-cost communication technique for monitoring with the least-cost technique for manipulation. For example, according to Table 9.5, the leader of a large dispersed group will compare the cost of standardized impersonal monitoring with the cost of standardized face-to-face manipulation, and choose monitoring or manipulation according to which of these 'best of their kind' techniques affords the lower cost.

The relevant cost in the calculations is the overall transaction cost (expressed, as before, on a per capita basis). This cost includes not only the direct cost, but also the indirect cost stemming from residual cheating (or non-participation) that the technique cannot eliminate.

So far as direct costs are concerned, it was assumed in Chapter 2 (and thereafter) that only the fixed cost of any monitoring or manipulation technique varies with the size or dispersion of the group. The marginal cost of intensified manipulation was assumed to be constant throughout, whilst the marginal cost of increasing the size of hostage was set to zero.

Because the marginal cost of manipulation is constant, the manipulation strategy—and hence the crime rate—does not normally vary with the size of group. Thus the indirect costs of manipulation are constant and, with a uniform distribution of sensitivity, equal to the direct variable costs. (This result does not always hold—teamwork with independent probabilities (Section 7.3) has size-dependent manipulation and a size-dependent crime rate, for example. The result also needs to be amended when the role of personal reputation in small groups is taken into account, as indicated above.)

The indirect costs of monitoring, by contrast, are not only constant, but normally zero. This is because the zero marginal cost allows the hostage level to be set sufficiently high to eliminate all cheating. (If participation is voluntary, however, then indirect costs may become positive, as non-participation can reduce the overall efficiency of the group.) The variable costs of monitoring are zero too, for obvious reasons. Thus with compulsory participation the total transaction cost of monitoring is simply equal to the direct fixed cost.

Given these assumptions, Fig. 9.3 illustrates, in a highly stylized way, how the size and dispersion of the group will influence relative transaction costs. It is assumed, for simplicity, that the dispersion of the group increases along with its size so that the two separate determinants collapse into one.

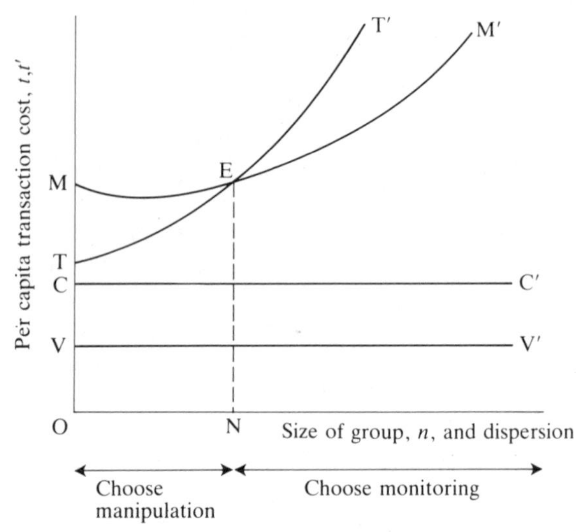

FIG. 9.3. Impact of size on the choice between monitoring and manipulation without personal reputation effects

In the absence of personal reputation effects, the direct variable cost of manipulation is typically independent of the size of group so that the variable cost schedule VV′ is horizontal. The indirect costs of manipulation are equal to the direct variable costs and are independent of size as well; thus the sum of the indirect and direct variable costs is given by the horizontal line CC′ which is twice the height of VV′, i.e. OC = 2OV. The total transaction cost of manipulation is obtained by adding the size-related direct fixed cost to CC′ to get the upward-sloping schedule TT′. As shown the slope of TT′ increases with respect to size, indicating that the difficulties of maintaining the face-to-face communication required for effective manipulation accelerate as size increases.

The corresponding transaction cost for monitoring, represented by MM′, comprises only the size-related direct fixed cost. It is assumed that this cost increases more slowly with respect to size because monitoring systems can substitute impersonal communication for face-to-face contact more easily. Since the slope is lower (and the curvature too), monitoring will always be selected if the cost of monitoring is less than the cost of manipulation when size is small. Thus to generate interesting results it is necessary to assume that the cost of small-group monitoring OM exceeds the cost of small-group manipulation OT. Given the continuity of MM′ and TT′, this guarantees an intersection E corresponding to a critical size ON. The minimum transaction cost envelope TEM′ indicates that manipulation will be chosen at sizes below ON and monitoring at sizes in excess of ON. Which strategy is actually adopted will depend on the returns to group size, which in turn depend on the technology of group activity and ultimately on the underlying activity that the group is supposed to perform. This is considered further in Chapter 14.

The introduction of personal reputation effects means that for manipulation both the direct variable costs, and the indirect costs, are lower than before in small groups, but unchanged in large ones. The effect is illustrated in Fig. 9.4. The new schedules V_1V_1' and C_1C_1' have lower intercepts $OV_1 < OV$, $OC_1 < OC$, and slope upwards, instead of being horizontal like VV′ and CC′. As a result, the schedule T_1T_1' representing the new transaction cost of manipulation, has a lower intercept $OT_1 < OT$ than the original schedule, and a steeper slope. This shifts the intersection with

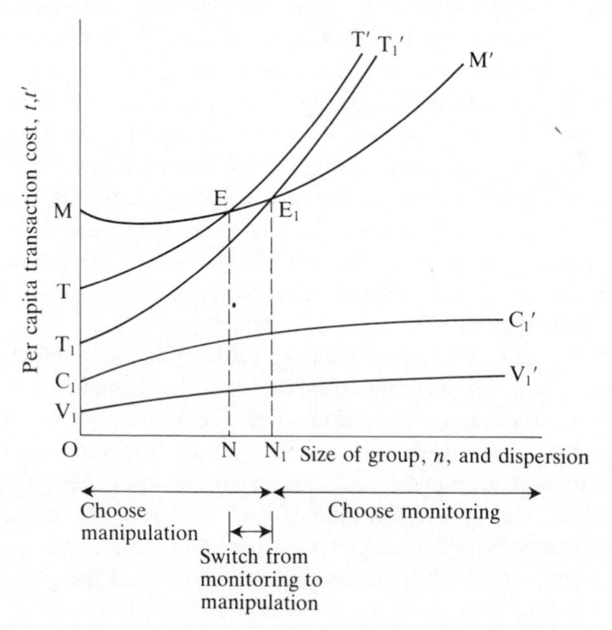

FIG. 9.4 Impact of size on the choice between monitoring and manipulation with personal reputation effects

MM′ to the right, from E to E_1, generating a higher critical size of group $ON_1 > ON$. This confirms the intuition that small group reputation effects increase the size of group to which manipulation can be efficiently applied.

9.6. Summary

This chapter has shown how the strength of reputation effects in small groups can improve their efficiency relative to large ones. One of the most fundamental effects of reputation is to allow the more sensitive individuals to identify the least sensitive individuals and avoid dealing with them. This reduces the risk that the viability of the entire group can be undermined by a high crime rate. This effect is of only limited value so far as a narrowly materialistic leader is concerned, however, because the main effect is to replace cheating with non-participation.

For reputation to be really effective individuals must recognize that their own behaviour has consequences for their reputation. It is only then that reputation reduces the incentive to cheat. When individuals expect to remain attached to the group for a considerable time, self-interest encourages them to defer cheating until later. If they always take the view that they are likely to remain for some further time, then cheating will be deferred indefinitely and honesty will prevail. Honesty is more likely to prevail the lower the individual's rate of time discount, the greater his optimism about other people's integrity, the larger the potential gains from encounters, the lower the one-off material incentive to cheat, and the lower his probability of leaving the group.

Under suitable conditions the leader can enhance this mechanism by announcing a prediction of total honesty. If everyone believes this then the prediction itself is self-validating: everyone will participate and no one will cheat. Because no one actually cheats, no one can be sure of how the reputation mechanism actually works. The important thing about the reputation mechanism, therefore, is that people should believe in it. This suggests another role for the leader, therefore—engineering beliefs that reputation mechanisms are effective. If the engineering is successful, the beliefs need never be put to the test.

The logic of the preceding argument depends directly on the stability of the group, but only indirectly on its size. Small size matters chiefly because of its effect on the efficiency of the reputation mechanism (or on the credibility of the leader's claim that it works well). The spatial dispersion of the group is important in this context too. Spatial dispersion inhibits face-to-face communication and so diminishes the effectiveness of moral rhetoric. As dispersion increases, monitoring mechanisms become more formal and standardized, but as they are relatively impersonal anyway this does not significantly impair their performance. Thus as group size increases, monitoring becomes relatively more efficient than moral manipulation. This is exemplified by the way that leadership by example, which can be very powerful in small and compact groups, can degenerate into the use of mass propaganda in larger and more dispersed groups.

If moral manipulation is to be employed successfully in large groups, leadership will normally have to be delegated, using intermediators of the kind discussed in Chapter 8. In large groups, therefore, the main alternative to formal monitoring is a highly decentralized type of moral leadership. This requires that the overall leader be effective in developing and educating other leaders who can translate his principles into practice in small-group situations.

PART IV

Collective Co-ordination and Social Justice

10

From Free-riding to Philanthropy

10.1. Introduction

The next three chapters focus on collective aspects of group activity. The preceding analysis has emphasized the role of individual efforts (Chapter 2), pairwise encounters (Chapters 3–6), and teams (Chapters 7 and 8). In each case group performance is determined by aggregating the value added by individual constituents of the group. The emphasis now switches to a more organic or holistic view of group activity. There is more concern with activities such as public-good provision and redistribution, where all the members of the group are involved together.

This chapter focuses on issues raised by the literature on public choice (Mueller 1990). It is this branch of economic theory, above all others, which has taken seriously the limitations of the legal system in co-ordinating group activity. It therefore provides a natural arena for displaying the potential benefits of moral manipulation in their strongest light.

The redistribution of income is an important group activity. It is shown later on in this chapter that redistribution raises problems closely analogous to those of public-good provision, and it is demonstrated that moral leadership is well suited to engineering redistribution that carries the general consent of the group.

An important aspect of justice concerns the distribution of income between the leader (and his élite) on the one hand and the rest of society on the other. To some extent competition for leadership constrains the abuse of the leader's power, but moral self-restraint by the leader is important too. These issues are explored in Chapter 11.

When the distribution of income is perceived as unfair, there is not necessarily unanimity about how equity should be restored.

If the problem is clearly the abuse of the leader's power then mass insurgency may result, as described in Chapter 11. A more common situation, however, particularly in democratic industrialized countries, is that the distribution of income between different occupational groups is regarded as unfair. This can lead to the emergence of competing factions. Individuals with similar economic interests collude, with the aim of enhancing their bargaining power against other factions. Each faction will normally have its own leader, so the role of the overall leader becomes that of 'holding the ring' while the leaders of the factions settle their differences. This reveals another negative side of leadership, besides the abuse of power—namely the ability of factional leaders to manipulate more effectively than the overall leader of the group. Part of the explanation may lie in the relative ease of moral manipulation in small, stable, and compact groups of the kind described in Chapter 9. Issues relating to collusive factions are considered in Chapter 12.

10.2. The Free-Rider Problem

The free-rider problem has a special place in economic theory because it highlights the difficulty of devising incentive-compatible contracts in certain kinds of situation (Buchanan 1965). The problem is usually discussed in the context of public goods although, in fact, it is more general than this. The essential feature of a public good is that, once produced, it has underutilized capacity. This means that one person's use of the good does not deny some other person use, as it does with a private good. This is a technological feature of the good. But it is linked, in many cases, with a contractual feature, which is that it is impracticable to exclude people from access to the good. This means that it is impossible for anyone to capture the value of the good by charging for access, since people cannot be denied access in the first place (Olson 1965). Economists perceive this as a serious problem because it means that the provision of public goods cannot be decentralized using the profit motive. The person who finances the production of the goods has no means of recovering his costs, so that a private-enterprise system will result in underprovision of public goods.

An enlightened entrepreneur might, under these circumstances, invite people to contribute to the cost of the good, on the basis that unless they contribute there will be no good for them to enjoy. But a self-interested individual has no incentive to contribute unless he believes that his own contribution will be the marginal one. If few other people contribute then his sacrifice is in vain, whilst if other people are willing to contribute he may as well take a 'free ride' on them. When he believes that everyone else is self-interested too, the likelihood that his contribution will be critical is very small indeed.

The root of this problem, as many writers have recognized, lies not in the technology of the good itself but the difficulty of monitoring its use. If users could be identified then an owner of the good could administer a charge. It is because monitoring is prohibitively costly that some alternative mechanism must be used.

A common arrangement is to finance public goods out of taxation—local services, such as road maintenance, are financed out of local taxation, and national services, such as defence, out of national taxation. The basis for taxation is that because of the difficulty of exclusion, everyone is a potential user, and so it is reasonable to overcome the free-rider problem by using the force of government to make everyone pay.

Tax systems bring problems of their own, however—income-related taxes, for example, distort incentives with respect to work and saving, and less distortionary taxes often incur heavy administrative costs. It seems natural, therefore, to invoke the response to monitoring difficulties which has been emphasized throughout this book—namely moral manipulation to make free-riders feel guilty.

10.3. Manipulating Contributors

The analysis of public goods requires a somewhat different approach from that of the previous chapters, because public goods are not typically provided as a result of random encounters but through a concerted effort involving all members of the group. It is therefore appropriate, instead of working with a probability p that other people will cheat, to work with the ratio j of the number of members who do not contribute to public-good provision to the total size of the group.

To maintain continuity with the previous analysis, it is assumed that each contributor provides effort at a cost $e > 0$. The value of the public good, it is assumed, is directly proportional to the number of people who contribute to its production. The marginal product of a contributor is $a > 0$. It should be noted that this refers to a constant-returns engineering technology and assumes that people who contribute do not slack (see Chapter 7). In practice many public goods exhibit scale economies in production but allowing for scale economies does not alter the analysis in any significant way. The public-good aspect of technology is captured by the fact that the material consumption of everyone in the group, whether or not they contribute, is equal to the value of the good. Failure to contribute incurs an emotional penalty $g \geqslant 0$, where g depends, as before, on the sensitivity of the individual and the intensity of manipulation.

The follower decides whether to contribute using the data shown in Table 10.1. He does not know the number of other people who will contribute, z, but this does not matter because z does not affect his decision. The net gain to contributing is equal to the material gain from the enhanced value of the public good, a, plus the emotional benefit of avoiding guilt, g, less the cost of effort, e. The follower contributes if the net gain is non-negative, and this implies a lower bound on the level of g:

$$g \geqslant e - a. \tag{10.1}$$

If $a > e$ then no incentive problem exists, but normally this condition is not satisfied.

TABLE **10.1.** *Follower's data set for public-good free-rider decision problem*

Strategy	Reward
Contribute (honesty)	$(z+1)a - e$
Not contribute (cheating)	$za - g$

Suppose, to begin with, that the leader is a narrow materialist, aiming to maximize consumption benefits irrespective of the effort involved. His objective expressed in per capita terms, is

$$v_1 = a(1 - j)n, \tag{10.2}$$

where

$$j = 1 - (z/n) \tag{10.3}$$

and n is the size of the group. He maximizes utility

$$u_1 = v_1 - c, \tag{10.4}$$

where c is the familiar cost of manipulation as given in equation (5.7).

With a uniform distribution of sensitivity and $e > 0$,

$$j = (e - a)/\theta. \tag{10.5}$$

Substituting (10.2), (10.3), and (10.5) into (10.4) and taking $\theta > 0$ gives

$$u_1 = an\{1 - [(e - a)/\theta]\} - c_f - c_v\theta. \tag{10.6}$$

An interior maximum satisfies the first-order condition

$$an(e - a)/\theta^2 = c_v, \tag{10.7}$$

whence

$$\theta_1^e = [an(e - a)/c_v]^{\frac{1}{2}} \tag{10.8.1}$$
$$j_1^e = [(e - a)c_v/an]^{\frac{1}{2}} \tag{10.8.2}$$
$$v_1^e = an - [an(e - a)c_v]^{\frac{1}{2}} \tag{10.8.3}$$
$$u_1^e = an - 2[an(e - a)c_v]^{\frac{1}{2}} \tag{10.8.4}$$

Under the assumed conditions the second-order conditions are always satisfied.

As noted earlier, narrow materialism generates the appearance of wealth within a group, but tends to produce excessive material output. Members are encouraged to provide too much effort from a broader point of view. A broad materialist would maximize instead

$$u = v_2 - c, \tag{10.9}$$

where

$$v_2 = (an - e)(1 - j). \tag{10.10}$$

Replacing (10.2) and (10.3) with (10.9) and (10.10) and reoptimizing gives

$$\theta_2^e = [(an - e)(e - a)/c_v]^{\frac{1}{2}} \tag{10.11.1}$$
$$j_2^e = [(e - a)c_v(an - e]^{\frac{1}{2}} \tag{10.11.2}$$
$$v_2^e = an - e - [(an - e)(e - a)c_v]^{\frac{1}{2}} \tag{10.11.3}$$
$$u_2^e = an - e - 2[(an - e)(e - a)c_v]^{\frac{1}{2}}. \tag{10.11.4}$$

The second-order condition

$$d^2u/d\theta^2 = -2(an - e)(e - a)/\theta^3 < 0 \tag{10.12}$$

is satisfied only if

$$an > e \tag{10.13}$$

which is the conventional condition (expressed in per capita terms) that the benefits of the public good exceed the cost.

Comparing equations (10.8) and (10.11), on the assumption that (10.13) is satisfied, shows that with broad materialism the intensity of manipulation is lower, and the crime rate (the proportion of non-contributors) is higher than before. Effort is conserved, but material consumption is reduced as a result.

Note that in both cases the intensity of manipulation increases with respect to the size of group, n. In the case of narrow materialism, for example, the intensity of manipulation is proportional to the square root of n. This is a consequence of increasing returns to the size of group which are implicit in the public-good technology. (These are quite distinct from returns to scale in production, which were assumed to be constant earlier.) The good is a public good because it is underutilized relative to its capacity (which could be infinite), and so the larger the group, the more efficiently the good is utilized. The more efficiently it can be utilized, the more important it is to see that it is produced, and hence the greater the incentive for the leader to manipulate. This result may be compared with the result for a private good in the next section.

10.4. Monitoring Problems with a Private Good

Not only is moral manipulation an important practical instrument, but studying the free-rider problem from the standpoint of moral manipulation provides new insights into the nature of the problem too. It indicates, for example, that the magnitude of the incentive problem—and hence the intensity of manipulation required to overcome it—is actually greater when the same type of monitoring problem exists but it is a private good rather than a public good that is involved. This is because when an individual contributes to the production of a public good he knows that he will have free access to it, whereas when he contributes to the production of a private good whose use cannot be monitored (sometimes called a common good) there is no such guarantee, and so the material component of his reward is even lower than it is in the public-good case. It turns out, though, that although the incentive not to contribute is greater, the resulting loss may not be so severe, so far as the leader is concerned, because the private good does not afford the same economy to wider utilization as does a public good.

Free-rider problems with private goods are nowhere near so common as free-rider problems involving public goods, but significant instances can still be found. Consider, for example, a community of fishermen concerned to conserve local fish stocks. Every member can, in principle, make a contribution to conservation by an expenditure of effort costing $e > 0$. Individual efforts have additive effects, so that each individual effort improves the potential annual catch for the community by $a > 0$ independently of whether similar efforts are made by other people.

If everyone devotes the same time to fishing, and has similar skill, then the total catch will, on average, be divided equally between members of the community. Because the fish circulate in local waters, there is no direct connection between conservation at a given spot and the number of fish caught there. There is, therefore, no simple way for a fisherman to capture the benefits of his own conservation efforts through his fishing strategy. Fishermen cannot be paid directly for their conservation efforts because, it is assumed, they are residentially dispersed around the coast and there is no durable evidence of their work so that their efforts cannot easily be monitored.

Under these conditions each individual benefits to only a small extent from his own personal effort. With a community size n, his own catch will increase, on average, by a/n, whilst the rest of the community, as a whole, benefits by $a(n - 1)/n$. In the absence of moral manipulation, therefore, personal effort is unlikely to be worthwhile.

The data for the decision problem are given in Table 10.2, in which z represents the unknown number of other fishermen who will contribute to the conservation effort. The condition for contributing to conservation is that

TABLE 10.2 *Follower's data set for private-good free-rider decision problem*

Strategy	Reward
Contribute (honesty)	$[(z + 1)a / n] - e$
Not contribute (cheating)	$(za/n) - g$

$$g \geqslant e - (a/n). \tag{10.14}$$

This indicates that an incentive problem exists whenever $e > (a/n)$, which, for given e and a, is a weaker condition than before. This confirms that incentive problems are potentially greater with private than with public goods.

A narrowly materialistic leader is concerned with the additional per capita catch generated by conservation,

$$v_1^\dagger = a(1 - j). \tag{10.15}$$

He maximizes

$$u_1^\dagger = v_1^\dagger - c, \tag{10.16}$$

subject to the usual cost conditions and the constraint

$$j = [e - (a/n)]/\theta, \tag{10.17}$$

which is derived in the same manner as before. An interior maximum satisfies the first order condition

$$a[e - (a/n)]/\theta^2 = c_v \tag{10.18}$$

whence

$$\theta_1^{\dagger e} = \{[a(e - (a/n)]c_v\}^{\frac{1}{2}} \tag{10.19.1}$$

$$j_1^{\dagger e} = \{[e - (a/n)]c_v/a\}^{\frac{1}{2}} \tag{10.19.2}$$

$$v_1^{\dagger e} = a - \{a[e - (a/n)]/c_v\}^{\frac{1}{2}} \tag{10.19.3}$$

$$u_1^{\dagger e} = a - 2\{a[e - (a/n)]/c_v\}^{\frac{1}{2}}. \tag{10.19.4}$$

These results can be compared with the corresponding results (10.8) for a public good, provided $e > a$. It is readily established that for $n > 1$ the optimal crime rate is always higher for a private good than for a public good. This is because the efficiency losses due to the free-rider problem tend to be lower with a private good. The results for optimal manipulation are more subtle however: for

$$n > a/(e - a) \tag{10.20}$$

manipulation is more intensive with a public good, whereas for small values of n manipulation is more intensive with a private good. Intuitively, the reason is that with a public good the leader's valuation of output increases with group size. With a private good the valuation does not increase with group size. But the incentive to free-ride is always greater with a private good. Thus when group size is small, and public good economies are consequently insignificant, manipulation is more intensive with the private good because of the relative severity of the incentive problem. Conversely, when the group is large the intensity is greater with the public good because the value of additional output is extremely high.

Results for broad materialism may be derived by a similar method. A direct comparison between the results for public and private goods is difficult, however. If the productivity of effort, a, is the same in each case, then a broad materialist will not wish to encourage production of the private good except when productivity is so high that there is no incentive problem at all for the public good. The comparison is therefore essentially trivial. The broad materialist undertakes no manipulation with a private good under the conditions where he undertakes positive manipulation with the public good. Under these same conditions, no one contributes to the private good although a high proportion of the group may contribute to the public good.

10.5. Philanthropy

Another obvious application of moral manipulation is to explain philanthropy. Philanthropic behaviour can, of course, be explained by introducing vicarious elements into the individual's utility function— for example, by giving a positive weight to other people's consumption (Collard 1978; Margolis 1982). Moral manipulation, however, provides a simpler explanation of the same phenomenon, and one which is potentially more satisfying, because it considers how these positive weights may have been generated in the first place.

Philanthropy is most naturally introduced as a limiting case of the situations discussed above, in which the contributor expects to gain nothing for himself, in a material sense, for a sacrifice he makes. A simple example is one in which a basically healthy population contains a single individual who is seriously ill. The leader is concerned to maximize the quality of care provided for this individual, and this is directly proportional to the number of people who are willing to give up time to nurse him. The material productivity of each nurse is $a > 0$. If the group is redefined to include only the healthy members then it is clear that no member of the group stands to gain materially from his own effort. Their only benefit is freedom from guilt or—looked at more positively— the feeling of satisfaction from caring for the sick.

This situation is so straightforward that its analysis is quite trivial. Since there is no material reward from nursing the sick person, the critical value of guilt required to sustain a contribution is given by the cost of effort, e. Since this critical value is independent of whether other people contribute effort, the decision problem is formally identical to that of the transactor in Chapter 3 confronting a PD over whether to cheat at trade. In that case too the incentive to cheat is fixed independently of other people's actions. The solutions given there apply in the present case as well. With a uniform distribution of sensitivity the optimal intensity of manipulation for a leader concerned only with the material well-being of the invalid is

$$\theta^e = (ae/c_v)^{\frac{1}{2}}, \tag{10.21.1}$$

and the corresponding crime rate is

$$j^e = (ec_v)^{\frac{1}{2}}. \tag{10.21.2}$$

The intensity of manipulation, according to (10.21.1), is directly proportional to the square root of the productivity of philanthropic effort and to the square root of its cost, and inversely proportional to the square root of the marginal cost of manipulation. The intensity of philanthropic activity, as measured by $1-j^e$, is shown by (10.21.2) to vary directly with its productivity, and inversely with both its cost and the marginal cost of manipulation.

The close analogy with the trader's PD indicates that the nature of incentive problems is often much the same whether random encounters or concerted group activities are involved. Indeed, the present example could easily be reformulated in terms of encounters by supposing that any member of a group is liable to come across someone else in need of help. Such help is difficult to provide through contracts, because there may be no witness to an on-the-spot agreement, and in any case the party needing help would be obliged to negotiate under duress. Advance arrangements involving, say, injury insurance are feasible, but often costly to make. Under these circumstances the moral engineering of philanthropy may be the most efficient solution to the problem. Market economies cannot improve on the 'Good Samaritan' for the provision of certain types of care.

10.6. Summary

This chapter has considered various situations in which the material reward to an individual does not correspond to the social value of his action (as perceived by the leader). When an individual faces a discrete choice between finely balanced alternatives this discrepancy will lead to too little socially beneficial action. The distortion of choice from the social ideal is caused by externalities—some of the benefit accrues to people who are not directly involved in the decision.

If monitoring were costless and all kinds of contract readily enforceable then the individual could receive compensation from the beneficiaries by threatening to exclude them. Alternatively the beneficiaries could promise to reward the individual for his beneficial action. But if external monitoring of individual activity is not feasible, or contracts are unenforceable in law, distortion of incentives cannot be corrected in this way. Fortunately, however, moral manipulation can 'internalize' the externalities by providing emotional rewards for beneficial actions and emotional penalties for selfish actions. The value of the net emotional reward of the marginal individual is equated to the material benefit his action affords to the rest of society. Moral manipulation, in fact, 'internalizes' externalities in a much more literal sense than do contracts, for the correction of incentives is actually internal to the individual—it is his conscience that resolves the externality problem.

It has been shown that voluntary redistribution raises an extreme form of externality problem. When redistribution is voluntary the initiative normally lies with the donor rather than with the recipient. In material terms the donor, who is the key decision-maker, actually loses, while the recipient, who is essentially passive, gains. Redistribution is not conventionally regarded as an externality problem because redistribution does not satisfy the Pareto welfare criterion used by most economists, which requires that everyone be better off. Redistribution, of course, leaves the donor materially worse off. In this book, however, the leader's preferences act as a social-welfare function, and so if the leader is altruistic redistribution may well represent welfare improvement.

One way of effecting redistribution is to reallocate property rights. This is the approach favoured by people who emphasize political change as an instrument of redistribution. It is reflected in Sen's view (Sen 1983) that lack of entitlements is a major cause of starvation in poor countries. It is also the approach of militant minorities in developed countries who claim greater freedoms as a legal right.

An alternative view is that because rights are ultimately moral rights, if they are widely recognized they may not need to be codified in law. Even if they are written into the law, the role of law may be mainly as a symbolic restatement of the dictates of conscience. A leader who emphasizes the moral rather than the politico-legal case for rights may be able to create a sufficiently high degree of moral unanimity that spontaneous philanthropy becomes widespread. The rights of minorities and of the poor may be upheld through moral legitimacy and without legislation. By promoting unanimity, moreover, the leader can prevent the factionalization of society that might otherwise occur when minority groups lobby to legalize their conflicting demands on the limited resources of society.

The examples discussed in this chapter concern only selected topics in an immensely wide field. Problems with monitoring, and with defective systems of property rights, create a wide range of externalities between individuals which no amount of tinkering around with material incentives can easily resolve. But whenever there is an externality of any kind there is the potential for a leader to manipulate a solution to it. This applies to public goods and public bads, to private goods and private bads, to localized encounters and to collective activities. It applies whether the externality is unilateral—for example, between donor and recipient—or mutual—as in the encounters discussed in earlier chapters. Wherever monitoring systems fail, manipulation is, in principle, available to fill the gap.

11

Distribution and Justice

11.1. Competitive Constraints on a Selfish Leader

An obvious criticism of the analysis of leadership presented in this book is that it adopts a romantic and idealistic view of the subject. Far from being examples of altruism and morality, it could be claimed, most leaders are actually quite selfish. Their rhetoric is pure hypocrisy—a smokescreen designed to obscure their selfish acts from public view. There is reason to believe that contests for power amongst potential leaders are often won by the most ruthless, and hence the least suited for the role. Even if the leader is honest when he comes to power, he may easily be corrupted by the power at his disposal. He may employ his power to make it difficult for anyone else to challenge him.

In the analysis of intermediation it was, indeed, assumed that those delegated by the leader would behave selfishly. But still the assumption was retained that the overall leader himself was altruistic. The selfishness of the intermediators was more crucial in some cases than others, however, because where competitive forces were active, the intermediators' autonomy was restricted by the need to match the terms that could be offered to the followers by potential rivals.

Competitive forces can apply to the overall leader too, and they have the same tendency to mitigate the consequences of a leader's selfish motivation. The focus here is on competition from other potential leaders within the group (competition between groups is considered in Chapter 13). The leadership contenders may be the heads of organized parties or factions, or simply populists able to manipulate hysteria in a crowd.

The effectiveness of internal competition depends very much on the degree of decentralization of information within the group. If, for example, only modest powers are delegated to intermediators, and they are monitored rather than manipulated, then information will be heavily centralized with the leader himself. The incumbent leader can use these established mechanisms to monitor political rivals, and employ his monopoly of key information channels to undermine their credibility through hostile propaganda. By contrast, in a strongly decentralized economy where independent intermediators are manipulated rather than monitored, the monitoring and misrepresentation of political rivals is much more difficult for the leader. Followers are exposed to a plurality of views and the leader must retain their trust through the plausibility of his arguments. This confirms, from a rather different vantage point, the familiar proposition that political freedom is most likely to be sustained in a decentralized market-oriented economy.

11.2. The Psychological Need for Justice

Competition is a process, and the process of internal competition requires rules. Without some rules, however primitive or informal, the winner of a competitive struggle for leadership cannot morally legitimate his victory. In principle the rules for competition are set by the followers, since it is only by their common consent that the leader can rule effectively. If followers are totally selfish there is unlikely to be much agreement between them. Everyone will favour the process and the outcome that give them the best material deal.

To achieve a modest degree of consensus, followers may have to invoke the concept of justice. Although the concept is riddled with ambiguities, it appears to be a key intuitive element in followers' thinking about the overall legitimacy of group activity. So far we have said little about the intellectual content of moral rhetoric, but it is evident that in most moral rhetoric, appeals to justice play a leading role (Hamlin 1986; Moore 1978). There appears to be a strong psychological need to believe that the group one belongs to is essentially a just one. The fact that this need is almost universal makes an appeal to justice a potent unifying force.

When justice is applied to a social process the requirement for *universal consultation* plays an important role (Vanek 1971). Since direct consultation between the leader and every follower is not really viable in a large group, a system of representation is widely used.

Moreover, since discussion may have to be very protracted if consensus is to be reached, deadlock-breaking mechanisms such as voting under majority rule may be employed. This can apply both to followers electing a representative and to representatives forming a collective view. Belief in the justice of consultation is often so strong that those consulted are willing to commit themselves publicly to the majority view even though they may privately have reservations about it.

Where outcomes rather than processes are concerned, the concept of *distributive justice* is crucial (Kolm 1969). Distributive justice has both a horizontal and a vertical dimension. The horizontal dimension refers to the equal treatment of those of equal status, whilst the vertical dimension refers to appropriate differentials in material consumption between those of different status. When followers are appraising their leader they are likely to recognize a difference in status, and so it is the vertical dimension that is crucial. When a leader is under suspicion, it is the relation between his own rewards and those of a typical follower which will come under closest scrutiny. This is not to say, of course, that if the leader has demonstrated favouritism, and so violated horizontal equity, this will not count against him too.

The concept of justice can work two ways so far as leadership is concerned. The conspicuous pursuit of fair policies can be a unifying force that powerfully strengthens the leader's position. The persistence of injustice can equally unify opposition to the leader. Even in a repressive regime, injustice can arouse people to anger and thence to revolt. As noted in Chapter 6, anger is extremely contagious. Contagion can be accelerated by mass gatherings, and such gatherings provide a forum in which populists, who may have few enduring leadership skills, can mobilize opposition to the point where followers are prepared to sacrifice their lives to dissipate their anger against the incumbent leader. In the light of this, it seems reasonable to postulate that the need for distributive justice is a significant constraint on incumbent leaders under a wide variety of circumstances. Even a highly repressive leader may find himself subject to an internal competitive constraint from the leader of a mob.

11.3. Distributive Justice

The way distributive justice constrains a selfish leader may be illustrated by an example based on the motivation of achievement discussed in Chapter 2. The simplest way to illustrate the point is to compare an altruistic narrowly materialistic leader with a selfish leader interested only in his own material consumption.

It is assumed that the selfish leader appropriates his consumption through a tax. When the leader relies exclusively on manipulation rather than monitoring, as in the case considered here, the tax cannot be related to the level of individual effort because individual effort cannot be observed. Neither can effort be inferred from individual output because that cannot be observed either. Any attempt to use such a tax would simply result in the underreporting of output. (Of course, an ethic of integrity could be introduced, supplementing the ethic of dedication, in order to encourage honest reporting, but that is not considered here.)

Since the tax cannot be related to the level of effort or output, it may as well be a lump-sum tax. A lump-sum tax can also be recommended on grounds of horizontal equity—it does not discriminate between followers of similar status.

Vertical equity requires that the tax burden cannot exceed a fixed share s $(0 < s \leqslant 1)$ of the average output. It is assumed that the average output is known to both leaders and followers. This is a reasonable assumption, given the assumption in the earlier chapters that the crime rate, which measures the overall incidence of slacking, is public knowledge. Knowing the crime rate, the output of a dedicated worker, $y > 0$, and the marginal productivity of effort, $\Delta y > 0$, everyone can calculate per capita output

$$v_1 = y - q\Delta y. \tag{11.1}$$

With the per capita lump-sum tax set at $t \geqslant 0$, vertical equity then implies

$$t \leqslant sv_1. \tag{11.2}$$

In the absence of the constraint (11.2), the only constraint on taxation would be a subsistence requirement. In the simple case where the subsistence requirement is independent of the level of dedication that the worker shows, the requirement that all workers (including slackers) can subsist is

$$t \leqslant y - \Delta y - w^*, \tag{11.3}$$

where w^* is the subsistence wage. It is assumed that (11.3) is a weaker constraint than (11.2), so that distributive justice becomes a constraint before subsistence does. Thus the cost to the leader of the distributive constraint, in terms of the reduction of tax required, is

$$\Delta t = (1 - s)y - (1 - sq)\Delta y - w^*.$$

Tax revenues cannot be used exclusively for the leader's consumption. They must cover the costs of manipulation too. Thus the selfish leader maximizes

$$u_5 = t - c, \tag{11.4}$$

where c is the familiar per capita cost of manipulation given by (2.11) and (5.7).

Table 11.1. *Data for follower's choice when subject to a lump-sum tax*

Strategy	Rewards		
	Material	Emotional	Total
Dedication	$y - e - t$	0	$y - e - t$
Slacking	$y - \Delta y - t$	g	$y - \Delta y - t - g$

The imposition of a lump-sum tax does not affect the incentive to slack, as the modified rewards shown in Table 11.1 indicate. Thus the criterion for dedication is the same as that implied by Table 2.1. With a material cost of effort $e > \Delta y$, the rule is to be dedicated if

$$g \geqslant e - \Delta y. \tag{11.5}$$

It follows that with a uniform distribution of sensitivity the crime rate is given by (2.13):

$$q = \begin{cases} 1 & e - \Delta y > \theta \geqslant 0 \\ (e - \Delta y)/\theta & \theta \geqslant e - \Delta y > 0. \end{cases} \tag{11.6}$$

Substituting (11.1) and (11.6) into (11.4), taking (11.2) as binding and setting $\theta > 0$, gives

$$u_5 = sy - [s\Delta y(e - \Delta y)/\theta] - c_f - c_v\theta. \tag{11.7}$$

The necessary condition for an interior maximum is

$$s\Delta y(e - \Delta y)/\theta^2 = c_v, \tag{11.8}$$

whence

$$\theta^e = [s\Delta y(e - \Delta y)/c_v]^{\frac{1}{2}} \tag{11.9.1}$$

$$q^e = \{[(e/\Delta y) - 1]c_v/s\}^{\frac{1}{2}} \quad (11.9.2)$$

$$v_1^e = y - [\Delta y(e - \Delta y)c_v/s]^{\frac{1}{2}} \quad (11.9.3)$$

$$u_5^e = sy - 2[s\Delta y(e - \Delta y)c_v]^{\frac{1}{2}} \quad (11.9.4)$$

The second-order condition for a maximum is

$$\mathrm{d}u_5^2/\mathrm{d}\theta^2 = -2s\Delta y(e - \Delta y)/\theta^3 < 0$$

and is always satisfied given that $e > \Delta y$, as assumed earlier.

Comparing these results with equations (2.15) shows that the altruistic leader corresponds to the case $s = 1$. In other words, if the selfish leader could appropriate all of the output for himself then the intensity of manipulation, and the consequent crime rate, would be the same as if he were altruistic. Of course, the implications for the followers would be very different since they would consume none of their own output.

For $s < 1$ the selfish leader manipulates less than the altruist because he is only concerned with the share of the productivity gains that accrues to him. As a result, the incidence of slacking is higher and average product lower. On the face of it, this suggests a familiar kind of equity–efficiency trade-off. The tighter the equity constraint, the less the leader manipulates, and so the lower is productive efficiency.

It is certainly correct to assert that in a group with a selfish leader, intense concern over equity will impair material performance. But the overall moral is nowhere near as simple as that. To begin with, because redistribution is effected through taxation, the results also imply that a high tax share will be associated with high output. The results do not therefore support the traditional liberal view that high taxes discourage effort and so reduce output. On the contrary, a high tax share encourages the manipulation of greater effort and so raises output. The difference arises, of course, because the present analysis ignores the conventional disincentive effects of taxation by assuming a lump-sum tax. Conversely, conventional analysis ignores the impact of taxation on the leader's incentive to manipulate, which is at the heart of the present analysis.

Although popular concern with distribution reduces average output, it actually increases followers' welfare—and it does so in two quite distinct ways. First, and most obviously, it prevents the leader from appropriating a large share of the output, and so leaves more for the followers to enjoy. But it also discourages the leader from the excessive manipulation engendered by his disregard of the material cost of effort. Recall from Chapter 2 that the narrowly materialistic leader, though altruistic, induces too much dedication by ignoring the cost of effort, and the selfish leader, for different reasons, does the same. By restricting the leader's tax share, followers counteract the tendency to over manipulation, and so help to restore a broadly materialistic or utilitarian optimum. In fact, from a utilitarian point of view the leader's tax share should be restricted to zero. This would eliminate any incentive to manipulate and allow everyone to slack. Since, by assumption, $e > \Delta y$, this is, from a utilitarian standpoint, what the followers would most prefer to happen. The overall conclusion, therefore, is that a distributional constraint may be a useful second-best strategy in a group under the control of a selfish leader.

Finally, it should be noted that in Chapter 2 there was no discussion of how an altruistic leader could appropriate sufficient resources to cover his costs. Lump-sum taxation, of the kind discussed here, provides an obvious solution. If the altruistic leader sets taxes just high enough to cover the costs of his optimal manipulation strategy given by (2.15) then the tax share will be

$$s^e = c(\theta^e)/v_1^e = (c_f + l)/(y - l), \tag{11.10.1}$$

where

$$l = [\Delta y(e - \Delta y)c_v]^{\frac{1}{2}}. \tag{11.10.2}$$

Because the intensity of manipulation favoured by a narrowly materialistic altruist can be quite high, it is possible that this share may fall foul of the distributive constraint.

This possibility arises, of course, because the followers do not distinguish between taxes which finance manipulation costs and taxes which finance the leader's own consumption. Where this distinction employed in the distributive equity criterion then there would be no problem. In practical terms, however, the present approach has much to recommend it, since it is quite possible that a well-meaning but misguided leader financing large amounts of propaganda may incur the opprobrium of his followers because of the tax burden involved.

11.4. Disincentive Effects of Taxation

It should not be inferred from the previous example that the distributive justice constraint will always be binding on a selfish leader. This happened to be true in the previous example because increased taxation created no disincentive to effort—indeed quite the reverse. The disincentive effect was missing because, although the leader could not monitor effort or output, he had sufficient power to implement a universal lump-sum tax. Under certain conditions, however, the leader may be able to tax only those who assemble at particular places, in order to participate in activity there. In a trading economy, for example, the leader may be able to tax only people who come to market, and not those who stay at home. This situation was quite common in medieval times. Similarly in a production economy the leader may be able to tax only those who affiliate to a team and not those who work on their own. This creates the familiar problem that high rates of tax erode the tax base both by discouraging participation and by encouraging those who do participate to do so unofficially through a 'black economy'. The leader therefore confronts a 'Laffer curve' in which, beyond a certain point, tax revenues decline as the rate of tax increases. If the maximum tax revenue, as a proportion of the output associated with the corresponding rate of tax, is less than the critical level s introduced earlier, then the distributive justice constraint will not be binding. It should be noted, however, that while the outcome may be just in vertical terms, horizontal justice is likely to be impaired because those who participate, and disclose their productive activities, pay tax whilst others do not. A system which rewards the dishonest and the non-participant at the expense of others is therefore objectionable on alternative grounds.

11.5. Contractarian versus Holistic Morals

Distributive justice has so far been discussed in isolation from the morality of cheating. It has been implicitly assumed that the rights and wrongs of dedication versus slacking can be considered independently of the underlying justice of the system in which work is carried out. Similarly, cheating on transactions has been considered independently of the degree of distributive justice implied by the terms of trade.

This separation of issues reflects a contractarian approach in which the obligation to honour the terms of a contract is independent of the distribution of rewards. The distribution of rewards may affect the material incentive to cheat, as indicated in Chapter 8, but it does not affect the morality of cheating, nor the emotional penalty incurred.

An alternative approach is to regard the moral system as a 'seamless robe' that envelops all human activity. Moral doubts about one aspect of the system raise questions about every other aspect too. Within this holistic approach, injustice in distribution may undermine the legitimacy of the distinction between honesty and cheating. Cheating the 'exploiter' may even become a noble calling, inverting the morality of the contractarian approach.

The holistic approach may also raise doubts about the institution of private property itself—an institution on whose existence all the preceding analysis is implicitly based. If property becomes 'theft', then the whole morality of contracting is overturned. Cheating is no longer a question of depriving other people of their legitimate property but of reappropriating resources that belong to the community.

From a philosophical point of view this holistic approach may have much to recommend it. Arguments based on 'natural law', for example, suggest that the case for cheating those who negotiate by placing their partners under exceptional duress cannot be lightly dismissed. From a practical point of view, however, the holistic approach is fraught with dangers, for it means that perceived injustice can render ineffectual all of the moral mechanisms discussed above. It becomes impossible to trust anyone who believes they have been treated unfairly. As the unjust leader's costs of manipulation increase dramatically, the crime rate rises and the economy plunges to catastrophe as no one is any longer willing to participate with anyone else.

This phase of economic collapse often seems, in practice, to precede a full-scale revolution of the kind threatened above (Tainter 1988). The collapse encourages revolution because the leader's power to counter-attack is perceptibly reduced. The fragmentation of society induced by the rising crime rate also encourages the bottom-up development of small self-help groups who can play an important tactical role in the revolt.

If the phase of collapse persists for some time before the revolution occurs then it may, unfortunately, cause further delays after the revolution before the economy picks up again. The atmosphere of mutual distrust may take a considerable time to dispel. The new regime may find it difficult to restore co-ordination until the justice of the new order is clearly demonstrated, for otherwise cheating will continue as it did under the old regime. This leads to a 'catch-22' situation, for redistribution may be difficult to effect until economic performance is improved, and individual's performance will not improve until justice has been restored. The contractual approach is again more flexible than the holistic approach, because it supports the ethic of integrity independently of the question of justice and so allows performance to improve before justice is restored.

Holistic morality therefore provides an economic trap for the unwary. A prudent leader will be wary about advocating a holistic approach. The message of this analysis is fairly clear—moral manipulation is best focused on specific virtues such as integrity, dedication, and loyalty (see later), whose value is independent of distributive justice. This message is, however, not particularly profound, because it merely reflects the materialistic leadership criteria on which the analysis is based. Those who favour implementation of the holistic approach, and reject the present analysis as morally naïve and culturally-specific, can maintain their position quite consistently, provided they recognize the probable economic penalties involved.

11.6. Asymmetry of Power and the Morality of Self-Discipline

The power that a leader enjoys with respect to his followers is one manifestation of a more general asymmetry that can affect relations between followers too. Asymmetric relations between followers have so far been ignored because of the simplifying assumption that all followers face similar material incentives. In practice differences in age, gender, intelligence, personal wealth, and so on create many situations in which the strong follower confronts the weak. The strong party in an encounter can inflict far more damage on the weak party than the weak party can inflict on him.

Concern over distributive justice motivates a particular form of moral manipulation here—an ethic directed specifically at the strong and calling for them to exercise special self-restraint against the weak. This ethic can be directed at the leader himself.

Established leaders can arrange for young potential leaders to be educated in this ethic. Victorian public schools in Britain are often credited with developing a cult of the Kipling-type leader in which the exercise of natural personal authority is held in check by austere self-discipline (Simon 1972; Vance 1985). There are parallels too in the ethic of simplicity of life-style and self-effacement which are found in other cultures. The obverse of this is the popular support for the underdog who confronts the unacceptable face of modern leadership—the bureaucrat who carries out inappropriate instructions in an inflexible way.

The ethic of the self-disciplined leader can reduce the need for reliance on representative processes and internal competition in securing distributive justice. The ethic strengthens the case for trusting the leader not to abuse his power and hence reduces the case for investing in expensive monitoring systems. The main difficulty with the ethic is that its effectiveness depends entirely on the personality of the leader and the moral education he has received. Unless the leader's own commitment to the ethic is so strong that any successor is almost certain to be under its influence too, it is dangerous for the leader to dismantle the monitoring institutions—even if there is no call for them under his own benign regime, they may well become crucial in the future.

11.7. Summary

This chapter has examined how the abuse of power by a selfish leader is contrained by the threat of internal competition. Leaders who decentralize responsibilities to intermediators face a relatively large number of internal competitors for the leadership role, compared to leaders who retain exclusive control for themselves. On the other hand, a leader who decentralizes may be able to engineer a competitive environment in which he can play off the contenders against each other.

The strength of popular support for rivals will depend on whether the leader's policies are perceived as socially just. Injustice may be associated with lack of consultation or the creation of a powerful clique of the leader's friends. Most fundamentally, though, injustice is associated with an inequitable distribution of income between the leader and the rest of the group. A model has been presented in which followers are aroused to revolt by excessive taxation, in much the same way that victims of cheating were aroused to vengeance in the model of trade in Chapter 6. The main difference is that in the present case anger is a group-wide phenomenon.

It is assumed that the leader knows the critical level of taxation which arouses revolt, and he optimizes the intensity of manipulation subject to this constraint. The leader, being selfish, chooses the intensity of manipulation to maximize the share of the tax takings available for his own consumption. It is shown that the intensity of manipulation is normally less than that of the altruistic leader described in Chapter 2. This is because a selfish leader is concerned only with that component of the material benefits of hard work that he can appropriate for himself. To some extent, therefore, followers who are concerned with social justice may impair the material performance of their own economy because they reduce the leader's material incentive to spur them to greater effort.

This is a special case of a more general problem—namely that excessive concern with social justice may undermine the whole basis of individual material incentives in the economy. Effective leadership is difficult to sustain when moral consensus is organized around notions of social justice. This can be a particular problem for those leadership contenders who win power through a popular revolution that has mandated them to establish social justice. Social justice is unlike the other virtues discussed elsewhere in this book—integrity, dedication, loyalty, and even optimism—all of which directly promote economic performance in one way another. Although it is true that notions of social justice may have a role in encouraging a leader to excercise self-restraint, it is almost invariably efficient to invoke the ethic of self-restraint directly, and simply encourage the leader to become more altruistic.

12

Collusion as Moral Crusade

12.1. Solidarity in Collective Negotiations

Collusion is an important form of group activity. It is usually concerned with redistribution. Members of one group typically collude against another group to increase their share of the total reward available. Very often the groups may be subgroups of a larger group. An internal struggle develops between two rival vested interests. Sometimes one of the groups may have just a single member. In the previous chapter, for example, the followers could have colluded against the incumbent leader in order to renegotiate the level of taxation. Such collusion, if successful, may peacefully resolve an injustice which would otherwise precipitate rebellion. If unsuccessful, however, it may simply mark the first step in the degeneration of the overall group into rival factions.

Collusion has certain similarities to team activity. Its effectiveness normally depends upon solidarity. Just as the output of a team is dictated by the strength of its weakest link (see Chapter 7) so the bargaining power of a collusive group is dictated by the self-restraint of its weakest member. Self-restraint is important because collusion provides enormous incentives to cheat. This is clearly evident in the case of a cartel—and most particularly when a boycott is used to increase cartel power (Harrington 1989).

A cartel can take many forms: a group of workers joining a trade union, a group of middlemen combining against their customers, and so on. When a cartel is established in a previously competitive situation, the output of each member is likely to be reduced as the customers reduce demand. As a result, each member will have excess capacity, at least in the short run. If the customer can easily switch between alternative sources of supply then it is well known that each member of the cartel has a strong incentive marginally to undercut the others in order to increase his market share. The incentive is greatest if no one else is thought likely to undercut as well. Thus the more honest other members are believed to be, the greater is the incentive to cheat.

If the customers refuse to pay the cartel price, then the situation may develop into a boycott. Alternatively, a producer's strike may be initiated in order to 'soften up' the customers. In either case the customers are deprived of their usual supplies and so their valuation of a marginal unit of the product rises. If substitute supplies are difficult to obtain then the marginal valuation may be very high indeed. This generates a quite exceptional opportunity to earn monopoly rents for any producer who is prepared to break ranks (provided no one else does so at the same time).

Given the strong material incentive to cheat, effective collusion requires a very strong monitoring mechanism. Monitoring is often very difficult, however, particularly where both producers and their customers are geographically dispersed. (When they are agglomerated, 'pickets' can be mounted or spies employed to check the movement of supplies.) An alternative to monitoring is, of course, moral manipulation by the leader of the cartel. Members may become self-monitoring if they believe in the legitimacy of their cause.

In this context the rectification of an injustice is a powerful motivation for solidarity in a cartel. A leader who can articulate the cartel's demands in moral terms may constitute a powerful unifying force. Forces of this kind have been used for many years by those trades unions who have related their own demands to a moral condemnation of the capitalist system within which they are obliged to operate (Booth 1985; Freeman and Medoff 1984; Naylor 1987). The generality of this moral mechanism is indicated, however, by the recent success that Catholic-dominated trades unions have enjoyed in the centrally planned economies of Eastern Europe. In the latter case it is the injustices of the one-party political system, rather than of the capitalist wage-system, that have provided the moral force.

12.2. Optimal Manipulation in a Cartel

Suppose that there is a fixed collection of resources that can be divided up in various ways between two groups. These could be the entire resources of a kingdom (as in conflict over taxation between a sovereign and his subject) or merely the surplus generated by some kind of production and trade, as in the case of unionization or a producer's cartel. The focus is on just one of the two groups. It has $n \geqslant 2$ members, while the other can have any number (including just one). The share of the total resources accruing to this group depends upon how well it can organize collusion against the other. To simplify the analysis the level of collusion in the other group is taken as given.

The structure of individual rewards is illustrated in Table 12.1. If all members of the group can be trusted not to break ranks then

TABLE 12.1. *Information set for member of a cartel*

Strategy	Other members	
	All honest	One cheats
Honesty (solidarity)	a	$a - b - d$
Cheating (breaking ranks)	$a + (n - 1)b - g$	$-g$
Perceived probability	$1 - P$	P

each member receives $a > 0$. If just one member cheats then he can obtain a much larger reward $a + (n - 1)\, b$, where $b > 0$. The application of a multiplier $(n - 1)$ to the material incentive to cheat b reflects the fact that cheating by one member redistributes income from all the other members of the cartel. The size of b reflects the amount of excess capacity available to the member, for if his capacity is small his ability to redistribute income from other members will be small as well. The remaining $n - 1$ honest members each receive $a - b - d$, where $d > 0$ (see Table 12.2). The negative coefficient on the asymmetric co-ordination gain d (see Chapter 3) indicates that cheating involves an overall net loss to the cartel.

TABLE 12.2. *Distributional effects of cheating in a cartel*

Per capita rewards	Number of cheats		
	0	1	2
Total	a	$a - [(n - 1)/n]d$	0
To cheats	—	$a + (n - 1)b$	0
To honest members	a	$a - b - d$	0

If two or more members cheat then, it is assumed, competition will drive down their rewards to zero. Their rivalry will be sufficiently conspicuous to become public knowledge. The cartel will be formally abandoned by the leader and so the honest are free to compete on the same terms as the cheaters. As a result, everyone obtains a zero reward.

It is often asserted that cartel cheating involves a PD, but in the present model this is not always the case. While it always pays to cheat when everyone else is honest, it does not necessarily pay to cheat if someone else is cheating as well. This is because, due to their own finite capacity, the inroad the cheat can make into other member's market share when others are holding the price firm may be fairly modest. Members may lose more if a price war is precipitated in which all the cartel rents are dissipated. The condition for matching a cheat to be worthwhile is

$$a - b - d < 0. \tag{12.1}$$

If (12.2) is not satisfied then the situation is not a PD. Instead of cheating always being best, the decision to cheat depends on the probability that another member will cheat as well. The higher is this probability, the lower is the expected material gain to cheating.

It is assumed that all members of the cartel regard the probability, p, that any given individual will cheat as very small. This implies that the probability, P, that someone else cheats can be approximated by

$$P = np, \tag{12.2}$$

and that the probability that two or more other people cheat is negligible. Thus the outcome where two people cheat is associated in members' minds only with the case in which one other person cheats and they respond by cheating themselves. This is the case in which they, and everyone else, obtain a zero reward.

Given a subjective probability p, and the material incentives described above, the level of guilt required to induce solidarity in an individual is

$$g^* = (n - 1)b - [a - d + (n - 2)b]np. \tag{12.3}$$

With a uniform distribution of sensitivity across the group, and an intensity of manipulation θ, the probability that an individual will cheat is

$$q = g^*/\theta. \tag{12.4}$$

Assuming that q is small, the probability that someone will cheat is approximately

$$Q = nq. \tag{12.5}$$

The top line of Table 12.2 indicates that an altruistic materialistic leader of the cartel will focus on the expected per capita reward

$$v = a - [(n-1)/n]dQ. \tag{12.6}$$

Maximizing utility

$$u = v - c \tag{12.7}$$

subject to (12.4) – (12.6) and the usual cost conditions gives

$$\theta^e = [(n-1)dg^*/c_v]^{\frac{1}{2}}. \tag{12.8}$$

To interpret this result it is easiest to take the special case $p = 0$ where all members are confident that none of the others will cheat. This is the case that provides the strongest material incentive to cheat, and so places the greatest demands on manipulation. Substituting for g^* using (12.3) and setting $p = 0$ gives

$$\theta^e = (n-1)(bd/c_v)^{\frac{1}{2}} \tag{12.9.1}$$

whence it follows that

$$q^e = (bc_v/d)^{\frac{1}{2}} \tag{12.9.2}$$

$$Q^e = n(bc_v/d)^{\frac{1}{2}} \tag{12.9.3}$$

$$v^e = a - (n-1)(bc_v d)^{\frac{1}{2}} \tag{12.9.4}$$

$$u^e = a - 2(n-1)(bc_v d)^{\frac{1}{2}}. \tag{12.9.5}$$

Equation (12.9.2) suggests that to interpret these results correctly it is necessary to assume that the marginal cost of manipulation c_v is very small. Otherwise the actual incidence of cheating will be high and the earlier assumptions will be invalidated. This requirement should not be overstated, though—the assumption that the probability of cheating is small exists first and foremost in the minds of the members and the leader, and it is only if they have regular experience on which to form their expectations that consistency demands that their beliefs be validated.

Equations (12.9.1), (12.9.4), and (12.9.5) indicate that manipulation becomes more difficult as the size of group increases. The intensity of manipulation increases linearly with respect to size and, as a result, the average reward to members and the leader's utility decrease linearly; a large incentive to cheat honest members, b, and a large asymmetric co-ordination gain, d, raise the intensity of manipulation and reduce rewards as well.

Equations (12.9.2) and (12.9.3) indicate that the incidence of individual cheating is independent of the size of group. This is because the intensity of manipualtion rises as group size increases to counteract the increasing material incentive to cheat. But with a constant incidence of individual cheating, the probability that someone cheats increases linearly with the size of group.

These results confirm the common intuition that the ability to collude is strongly influenced by the size of group. Small groups are much better adapted to collusion than large ones. It was noted at the outset that collusion has certain affinities with team activity, and the influence of size on the overall incidence of cheating within the group is one of them. It should be emphasized, however, that other aspects are crucially different. In a team the incentive to cheat is greatest when others cheat, whereas in a cartel the incentive to cheat is greatest when others are honest. The problem of size in a team is that as size grows the perceived probability of cheating increases, whereas in a cartel the problem is primarily that the material incentive to cheat increases as there are more members to redistribute resources from. These differences mean that size is potentially a far more crucial issue in a cartel than in a team. As a result, the hallmark of the small group, as opposed to the large one, is not so much that it is well adapted to teamwork as that it is a natural focus for collusion.

12.3. The Monitoring Alternative

The monitoring alternative is often not viable in the context of collusion. One reason is that individual cheating is difficult to detect—it is only when cheats find out that the customer is playing them off against other cheats, and their rivalrous responses attract the attention of others, that cheating may come to the leader's notice.

Another reason concerns the nature of the hostage that members must place if the leader is to have an effective means of punishment. In trade the goods being exchanged can be used as natural hostages by a leader who intermediates between the followers. In team production the members' share of the product can be held as hostage by the leader. But in the case of a cartel no such natural hostage exists. Members must be required to place special deposits with the leader if the hostage mechanism is to be used. These deposits, moreover, will have to be large, particularly in a large group, for the reasons indicated above. This creates a problem for the members, not only in financing their hostage payment (or 'bond') but in trusting the leader not to abscond with the very large sums involved.

It is because of these difficulties that leaders of cartels which rely on monitoring rather than manipulation have to invoke threats that honest members will collude against any cheats that are discovered, and punish them by spoiling the market through oversupply. Unfortunately such threats are not usually credible because in spoiling the market for others the honest members spoil the market for themselves as well. They are only credible (at least in a one-off encounter) if the leader can harness the anger of the honest members by arousing them to punish the cheats purely for the emotional benefits involved (see Chapter 6). But this brings the argument almost a full circle back to manipulation, for the strategic control of anger is an aspect of moral manipulation. If the leader can control members' emotions sufficiently for this purpose, he can probably build up an atmosphere of solidarity to begin with which will obviate the need to rely on threats of this kind. It seems therefore, that where collusion is concerned, moral manipulation dominates monitoring in almost all respects.

12.4. Redistribution between Factions

Economic liberals who are critical of group-centred activity often focus their criticism on the propensity of groups to engage in collusion. There is certainly substance to this criticism—the efficiency gains from team production or internal trade may be neglected as effort focuses on the redistribution of income from other groups. Collusion by one group may well induce defensive collusion by another group with which it trades. The worsening climate of intra-group relations may well prevent efficiency gains being realized from intra-group trade as well.

Because collusion is particularly effective in small groups, a small group may often be able to hold a much larger group to ransom. Although the large group may benefit from technological scale economies and a large internal market, it may be vulnerable to redistributive demands because the members who make up the group lack solidarity. Thus the small-group leader can play off different individuals in the large group, harnessing interpersonal competition in the large group as part of a 'divide and rule' policy. If the leader of the large group is unable to co-ordinate his members' dealings with the small group, the small group may become quite wealthy through redistribution even though its underlying productivity is quite low. Craft trade unions, and even nationalistic terrorist groups, exemplify small groups of this kind.

The inability of a large-group leader to constrain aggressive small-group leaders may encourage the formation of new small groups. Typically these will emerge as factions within the large group. In many cases the small-group leader will announce a general crusade to rectify injustice within the large group. In Chapter 11 the grievance was based on excessive taxation. But the present analysis highlights another possibility—that the injustice is actually that members of the new group do not enjoy the same consumption standards as members of other factions within the large group. It is not the violation of vertical equity that bothers them but the violation of horizontal equity instead. Their grievance is that they receive less than people of equal status in other small groups. This seems to be an important feature of the wage-bargaining process in heavily unionized economies.

It should be obvious to the leader of the large group that, when aggregated, the demands of these various factions cannot all be satisfied. This is because the norm for justice is usually set by the faction that has been most successful in redistributing income from other parts of the group. It is the process of collusion itself which needs to be tackled. Because moral manipulation holds the key to group solidarity, and solidarity holds the key to successful bargaining, the leader of the large group must either engage in battle with the leaders of the factions for the 'hearts and minds' of their members, or alternatively win over the leaders themselves. His moral message needs to contain a justification for unity and solidarity within the large group rather than a small one.

The ability to formulate such a message in convincing terms seems to be extremely scarce, and so the quality of leadership available effectively sets a limit on the viable size of a large group. A large group which forces its leader to overreach himself will degenerate into factions, and perhaps disintegrate completely before being reformed into a set of smaller groups of more manageable size.

Only a group of modest size, in which the leader can restrain the collusive propensities of subgroups through manipulation of the leaders and their members will be viable in the long run.

12.5. Loyalty

It was noted in Chapter 9 that self-interested integrity was most likely to emerge in a stable group, because with stable membership the expected frequency of encounters with other group members is relatively high. In the quest for integrity, therefore, a leader may attempt to discourage membership turnover, and one way of doing this is to build an ethic of loyalty.

Loyalty is of particular importance when the group is engaged in collusion against others. The other groups may stand to gain a great deal from 'poaching' members from their rival. They strengthen themselves and weaken their opponent at the same time. The opponent can be very seriously weakened in some cases—in the case of two armies disputing territory, for example, a defecting member may bring vital military intelligence with him.

The incentive governing the decision to quit a group and join a rival is very similar to that of cheating a fellow member of a cartel, and there is little to be gained by restating the analysis with only minor alterations. It is not difficult to show that engineering loyalty is much easier in a small group than a large group, and for reasons analogous to those which were indicated above. Loyalty is particularly important in small groups, because, as noted earlier, the reputation mechanism by which stability of membership reduces cheating is strongest in such cases.

Differences in attitudes to loyalty are, in fact, one of the more obvious features of international cultural variation. Casual comparison would also support the idea that a strong ethic of loyalty is associated with a militaristic tradition. Thus both Britain and Japan, which have a history of militaristic imperialism, have strong elements of loyalty in their culture. Unfortunately for Britain, however, loyalty in the industrial sphere has been focused mainly on factions based upon rival craft unions, whereas in Japan a greater degree of harmony has been achieved by organizing unions on a company basis. The question of loyalty to factions is considered further in the next chapter.

12.6. Summary

This chapter concludes the discussion of manipulation as a method of improving the co-ordination of group activity. By focusing on one of the less attractive aspects of co-ordination—namely collusion designed to redistribute income away from other groups—it provides an antidote to the rather idealistic picture of manipulation painted in the earlier chapters.

It has been shown that the moral manipulation of cartel members has certain similarities with the manipulation of team members described in Chapter 7. There is, however, the important difference that in a cartel the incentive to cheat is normally greatest when no one else is expected to cheat, whilst in teamwork the reverse applies. The magnitude of the material incentive to cheat when others are honest is also very much greater in the case of a cartel. Finally the material incentive to cheat increases significantly with size of membership in a cartel, whereas it does not in a team. This means that the intensity of manipulation is likely to be much greater in a cartel compared to a team, particularly when the number of members involved is large in each case. The incidence of cheating in a cartel will be correspondingly high—confirming a well-known result of conventional economics.

This conventional result is derived from an analysis of monitoring rather than of manipulation, however. Monitoring is normally even more difficult than manipulation where a cartel is concerned. Monitoring individual members is certainly more difficult in a cartel than in a team because in a typical cartel the team members are more widely dispersed. Whereas monitoring is an attractive alternative to manipulation for certain types of teamwork, it is rarely preferable to manipulation where cartels are concerned.

Loyalty to a group can also be analysed as a cartel-cheating problem. Loyalty is particularly important in sustaining the advantages that small groups draw from the strength of their internal reputation effects. The main problem with loyalty is that, if wrongly directed, it may merely strengthen commitment to rivalrous factions and undermine commitment to the overall group.

PART V

Synthesis

13

Business Culture

13.1. Introduction

Chapter 1 raised the 'big question' of how national culture impacts upon national economic performance. The short answer is that national culture can have a significant impact on performance if it reduces transactions costs within the economy. This point has been repeatedly emphasized in the preceding chapters. But this short answer is also a little naïve. For on closer examination a national culture is not always a clearly definable homogeneous entity.

It has already been noted that moral manipulation can be delegated to intermediators, and this raises the possibility that the culture promoted by intermediators may be somewhat different from that promoted at the national level. Middlemen engaged in retail trade may promote their own brand of culture, designed to be in harmony with consumption of their own 'brand' of good. This culture may be geared more to creating a monopolistic advantage in the product market rather than reducing transactions costs *per se*. Similarly, team leaders engaged in production may favour a strongly paternalistic culture which emphasizes the natural authority of the founder of the firm. The founder's claim to the personal allegiance of his employees may at times conflict with the claims of the national leader—if the founder wishes his employees to connive in illegal trade practices, for example. It is therefore important for the national leader to ensure that intermediators behave in a responsible way.

In a modern market economy the chairmen and chief executives of large firms have an influential leadership role. Because their firms typically integrate production and marketing, their leaders combine the roles of team leader—co-ordinating the efforts of individual production workers and line managers—and middleman—buying raw materials and components from suppliers and selling them on as finished products to final customers. These business leaders compete with each other for the allegiance of customers and employees. They may also compete with the national leader where conflicts of economic interest arise, but there is the possibility of a symbiotic relationship with the national leader too. This chapter investigates what system of moral values is most likely to achieve a symbiosis in which businesses prosper by promoting a morality which is aligned with overall national goals.

13.2. A Typology of Groups

A national economy comprises many groups of different kinds. The economic function of a group is to achieve co-ordination in a particular sphere of activity (Pollak 1985; Roberts and Holdren 1972). Each sphere of activity is associated with encounters of a particular kind. Five main motives for seeking an encounter have been discussed in previous chapters: they are summarized in Table 13.1. In some cases, of course, encounters can be harmful, while in other cases cheating cannot be contained at an acceptable cost; in such cases the group effects co-ordination by allowing its members to avoid encounters instead.

TABLE 13.1. *Functions of groups*

Reference Number[a]	Function	Chapter reference
1	Production	2, 10
	(a) Technologically independent members	2, 10
	(b) Team complementarity	7, 8
2	Trade	3–5, 8
3	Public good provision	9
4	Philanthropy	9
5	Collusion	11

[a] See Table 13.2.

Seven main types of group can be distinguished, as indicated in Table 13.2. They range from small units such as the family up to the nation-state itself. Three types of group are predominantly economic—the firm, the trade union, and the nation-state, while the others are mainly social—namely the family, the club, the religious group, and the local community. Different types of group have different functional specializations. The right-hand columns of Table 13.2 identify the major and the minor functions of each type of group. The relationships between different specialized groups are summarized in Fig. 13.1. The numerical column entries refer to the types of encounter listed in Table 13.1.

TABLE 13.2. *A typology of groups*

Type of group	Functions[a]	
	Major	Minor
Economic		
Firm	1*b*, 2	1*a*, 5
Trade union	5	3, 4
Nation-state	3	4
Social		
Family	1*b*	3, 4
Club	3	4
Religious group	3	4
Local community	3	4

[a] See Table 13.1.

The distinction between major and minor functions is indicative only. Thus some people might dissent from the view that the main functions of the family is to exploit gender complementarities in the rearing of children; they might prefer to emphasize the shared use of household public goods, such as consumer durables with underutilized capacity, or the philanthropy involved in caring for elderly relatives.

Several groups perform the same kind of function, but the type of encounter is rather different in each case. Several types of group engage in public-good provision, for example. While family provision of public goods is concerned with the continuous small encounters involved in eating and relaxing together, the club is concerned with intermittent encounters involving large groups of people pursuing specialized hobbies, and the local community focuses on the maintenance of infrastructure and the conduct of local rituals.

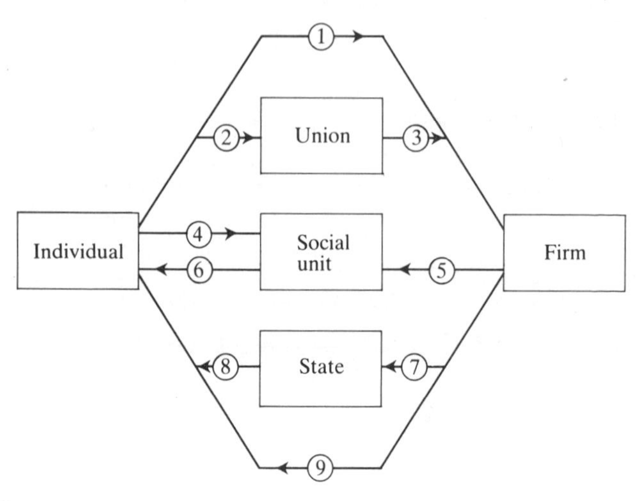

Key:

1 Individual non-unionized labour
2 Individual unionized labour
3 Collective unionized labour
4 Unpriced individual time and effort
5 Sales by firms to social units
6 Unpriced services supplied to individuals
7 Sales by firms to State
8 State services supplied to individual
9 Sales by firms direct to individuals

FIG. 13.1 Schematic diagram of some relations between different types of group

Each type of encounter raises its own incentive problems, and the nature of these problems will affect the type of ethic that is required. A trade union, for example, is geared to collusion, and so needs an ethic that emphasizes solidarity. Similarly, a club that is geared to the provision of public goods needs an ethic of selfless contribution to resolve the free-rider problem. The content of the ethic is a major determinant of the culture of the group.

In some cases the culture of one type of group may conflict with the culture of another. A trade-union leader, for example, may deny the legitimacy of private property, whilst the leader of a firm may emphasize the guilt associated with breaking contracts. In other cases, two cultures may be complementary. A business leader who encourages hard work through an ethic of dedication may also benefit families by encouraging dedication to family responsibilities too. Conversely, someone reared in a family with a strong work ethic (a farming family, for example) may make a dedicated employee.

The typical individual is a member of several different types of group. He is born into a family, joins a firm and a trade union, lives in a local community, and is a citizen of a State. He may also belong to a religious group and to various clubs (sports club, political party, charity, etc). If each one of these groups has a conflicting ethic, however, he may well become confused, and his commitment to all of the groups may suffer as a result. In some cases it can be resolved by breaking allegiance to one of the groups, but this can be very costly if it is, say, the family that is involved.

The national culture has a crucial role in reconciling different cultures. An important requirement of a national culture is that it acts as a unifying force. If it does not, the national economy may well degenerate into factions under the pull of localized collusion (see Chapter 12). Effective unification requires that the national culture reconciles—at least partially—the functionally specialized cultures of the different types of group.

A powerful national culture can in turn influence the intellectual environment in which the other groups operate. State propaganda can establish a cultural climate to which other groups must adapt if they are to survive. National culture can, therefore, be used to support the culture required by a certain type of group. A left-wing government may underwrite the values required by the trade-union movement, whilst a right-wing government may promote values that are in the business interest. In this way national culture can give a 'free ride' to certain groups by promoting their values; this not only reduces the group's costs of manipulation, but also enhances its general reputation by suggesting that its goals are in harmony with those of the State.

13.3. Competition for Allegiance

Different types of group are in competition for personal allegiance. Any individual has to allocate his time between the rival demands of family, employer, club activity, community service, and so on. The way he allocates his time will depend on the anticipated emotional rewards, as well as the material rewards, and the emotional rewards will in turn reflect the power of the leader's moral rhetoric in each group.

Competition for allegiance obviously exists between groups of the same kind too. But the nature of the competition is slightly different, because an element of exclusivity is involved. While individuals belong to several groups of different types they normally only belong to one group of the same type. This is particularly true of employment in a firm. It is also true of the family, the local community, the trade union, and the nation-state—the main exception being clubs, where some people may belong to several and others to none at all.

Because individuals belong to only one group of a given type at a time they face a discrete choice of allegiance rather than a continuous trade-off. In this choice at least one of the options is unfamiliar to them. While someone trading off time between family and firm generally knows the emotional benefits associated with each, someone choosing between their present employer and a new one has not already experienced the emotional benefits offered by the new one. Emotional benefits are more difficult to estimate than material ones; hence perception of these benefits will be strongly influenced by the reputation of the group.

Previous analysis in this book has concentrated on the reputation of the group with itself. The reputation of the group with outsiders—and in particular with potential recruits—has not so far been considered. The role of reputation in attracting good recruits gives all members a potential interest in maintaining it. But any member who damages the group's reputation by cheating will probably suffer less than will the membership of the group as a whole. There is, therefore, an externality problem in respect of external reputation—this is additional to the externality problem already analysed regarding the internal reputation of the group with itself. The effect of external reputation on the quality of membership raises the marginal return to reducing the average crime rate. This means that, so far as the leader is concerned, the optimal intensity of manipulation is increased.

Because group reputation serves as a public good, non-members may claim affiliation in order to gain a free ride. The damage that this can inflict is particularly severe when external reputation is important. The leader needs an identification system, such as a badge, by which members can be recognized. The badge may be quite subtle—the use of a special language or style of dress, or demonstrable familiarity with a special ritual.

Where affiliation is discrete, the same result can often be obtained by excluding members of other groups. Discrimination against non-members is effected using signals transmitted by the other groups's badging systems. So far as the leader is concerned, discrimination is an effective solution to free-riding when reputation operates at the group level. Conversely, the antidote to discrimination is to refine the manipulation mechanism so that it operates at the individual level instead.

13.4. Loyalty

An individual who quits a group to join another generally inflicts a loss on the group he leaves. The transaction cost incurred in replacing him cannot normally be passed on to the individual concerned. The loss is often increased by the abrupt nature of the departure, which in turn reflects the individual's belief that the benefits he derives from the group will decline if it is known in advance that he plans to leave. A person who plans to leave attaches less value to his reputation within the group because the anticipated number of future encounters is reduced and so, as indicated in Chapter 12, material self-interest is not so powerful a check on cheating. In so far as other members recognize this they will avoid dealing with the person concerned, and so his own optimal strategy is to leave right away.

The risk of quitting can be reduced by developing an ethic of loyalty within the group. The loyal individual feels guilty about leaving and so may stay even if the material rewards are poor (Hirschman 1970, 1976, 1982). The impact of loyalty will be strongest on the most sensitive individuals—who are generally also the most valuable (honest, dedicated, etc.) members of the group. The manipulation of loyalty is therefore a useful selection mechanism by which the most effective members can be retained and the least effective discarded through voluntary separation.

A similar selection mechanism is at work in recruitment. The more sensitive an individual, the greater is the weight, within a given package of benefits, that he attaches to emotional rewards. When manipulating a sensitive individual, the trade-off between material reward and intensity of manipulation is steeper than for an insensitive one. This allows a leader who offers intense manipulation but a low material standard of living to compete effectively against a leader who offers a better standard of living but lower manipulation.

So much is already implicit in the theory of non-pecuniary rewards in occupational choice. But what is particularly important is that the highly sensitive people who are attracted by the high-manipulation low-income policy are also the most responsive to the ethics of dedication, honesty, and loyalty. Once they join they will perform well, and will be disinclined to move even though they may be offered great material inducements to do so.

Two important consequences follow. The first is that because the most sensitive people are the most loyal, they are also the least mobile. This does not mean that they are geographically immobile (although loyalties to family and local community may have this effect), but that their initial commitment to a group has a strong element of irreversibility. They 'marry in' to their group. The best way to attract highly sensitive people is therefore to recruit them when they are making their initial choice. So far as firms are concerned, for example, this means that the most loyal employees are likely to be attracted as school-leavers or new graduates rather than by, say, head-hunting from rival firms.

The second point is that if sensitive people are to be encouraged to switch their allegiance, the recruiting leader may have to undermine the ethic of the other group in order to free the member from the guilt that he would otherwise feel. Unless the ethic is attacked, and the guilt destroyed, the cost of providing the leaver with material compensation for his guilt may be prohibitive. Such an attack can, of course, prove extremely painful for the target individual, who may face an emotional crisis as a result. A change of affiliation by a sensitive individual may well be linked to a vehement repudiation of the earlier ethic, if only to convince himself of his own liberation from guilt. At certain times, when the tide of general moral opinion changes, such dramatic switches by sensitive individuals can often be observed. It is also interesting to note that where religious affiliation is involved, those who profess liberation frequently forget the emotional satisfactions and portray the religion they have abandoned purely in terms of its teaching on guilt. The sense of loyalty is evidently more difficult to shake off than they anticipated.

13.5. The Culture of the Firm

In so far as economic theory has considered groups at all, it has concentrated on the organization of monitoring. In institutional economics, for example, managerial strategy and structure are usually assumed to be determined by the information requirements of monitoring rather than by those of manipulation. Switching the emphasis to manipulation provides a completely different perspective on the subject. The organization of manipulation is influenced by the type of ethic required, which in turn reflects the functions the firm is required to perform.

Table 13.2 shows that the firm stands apart from other groups in terms of the functions in which it specializes. It is not engaged primarily in public-good provision or philanthropy as several of the others are. But this is not because it is highly specialized—on the contrary, it is engaged in four types of activity. This is somewhat surprising, as the firm is often portrayed as a highly specialized entity created by the capitalist system.

The activities recognized in the table include the organization of team production (Alchian and Demsetz 1972) and the intermediation of trade. Wholesale and retail firms are pure intermediators, but even producers are intermediators in so far as they purchase inputs and resell them—albeit transformed and repackaged—as output (Casson 1982).

Production does not rely exclusively on teamwork, moreover. The manager of a firm may exploit indivisibilities in monitoring or manipulation to oversee technologically independent workers, or perhaps several independent teams. In other cases the firm may be exploiting a proprietary technology as an internal public good. The firm can then administer collusion amongst the teams or workers in order to appropriate a monopoly rent from the technology.

These varied activities mean that the firm requires a rich corporate culture for the purposes of moral manipulation (Hogg and Abrams 1988; Schein 1985).

1. The overall legitimization of the firm's activities rests on contractarian principles based on transferable property rights. It would be difficult for the firm to operate if private property rights were substantially incomplete or if cheating on contracts was condoned. A more holistic approach (see Chapter 10) can be tolerated within certain limits. But if, for example, distributive justice were considered more important than economic efficiency then some of the rents that accrue to managerial excellence would be difficult to justify.

2. The firm needs an ethic of achievement to motivate technologically independent workers, and a sense of team spirit, based on shared objectives, to prevent a weak link developing in a team.

The organization of the firm's internal trade in intermediate products benefits from an ethic of integrity, which is also important in promoting external reputation with customers and suppliers.

3. Collusion against customers will benefit from an ethic of solidarity. This solidarity must operate at the firm level, though, rather than at the level of the plant or team. The object is to prevent different plants from competing against each other and thereby dissipating monopoly or monopsony rents.

In a dispersed multi-plant firm, however, collusion may well develop at the plant level—particularly in plants that are small and compact, have low turnover (and hence stable membership) and which are relatively isolated (and hence communicate infrequently with headquarters on a face-to-face basis) (Church *et al.* 1989; Kerr and Siegel 1954). Successful collusion against headquarters by one plant may have a domino effect on other plants, leading to a proliferation of factions within the firm. An ethic of corporate unity is therefore required—an ethic which can only be effectively sustained by regular meetings between headquarters and subsidiary staff.

4. The contractarian ethics of the firm must not be so strong that management cannot understand the natural instinct for distributive justice that is articulated by trade-union leaders (Jaques 1982). Managers require an ethic of self-restraint and self-discipline if they are to avoid abusing their bargaining power in the labour market. By promoting justice spontaneously management may, indeed, unify the firm sufficiently to leave no role for the union—particularly if it respects workers' wishes for participation in the process of wage-fixing as well as the desire for a favourable outcome.

5. An ethic of loyalty is required to reduce turnover amongst staff. This is particularly important in respect of employees with valuable firm-specific skills (Hirschmeier and Yui 1981). Fortunately firm-specific skills are typically acquired through experience, and the greatest experience is accumulated by the most loyal employees. A firm can often tolerate a high level of turnover concentrated on short-serving employees; but if turnover begins to increase amongst long-serving employees then a crisis is imminent which calls for a revival of loyalty.

Managers may be tempted to promote loyalty to the firm at the expense of loyalty to other groups, such as the family or local community. If moral commitment to these groups is anyway very low, this tactic may be quite successful. But if the commitment is high, the firm presents the employee with a moral dilemma which he may be unable to resolve. In his confusion, commitment to all the competing groups may be reduced—including commitment to the firm. Loyalty should not therefore, as a rule, be inculcated to the exclusion of other groups.

A similar point applies to rival firms. Loyalty combined with solidarity may encourage a zealous hostility to other groups. This can damage the firm by inadvertently plunging it into economic warfare. The process of market competition has its code of ethics, like almost every other human activity, and deliberate violation of this code can make all competitors worse off (Knight 1935).

13.6. Two Styles of Business Leadership: Fear and Trust

A firm that relied entirely on monitoring could dispense with most of the values outlined above. It would not matter to the leader if people did not believe in the legitimacy of contracts provided they were willing to sign them out of economic necessity and could easily be fined if they defaulted. Problems of effort and integrity would be resolved by elaborate contingent contracts backed up by supervisory systems to spy on people. The difficulty of organizing collusion would be dealt with by lobbying for anti-competitive behaviour to be legalized. Turnover would be reduced by locking in key employees through personal loans and other devices.

There is strong evidence that successful firms do not rely exclusively on such methods (Ouchi 1981). Monitoring is almost never the firm's preferred solution to all incentive problems. Theory suggests two main conditions under which monitoring is most likely to be preferred—weak leadership and an unhelpful cultural environment.

A weak leader has high personal costs of manipulation. High marginal cost discourages intense manipulation, whilst a high fixed cost discourages any manipulation at all. Weak leadership may well be related to age and experience. A young leader may be emotionally immature, and unable to anticipate how others will respond to his rhetoric. He may, for example, assume that everyone shares his personal ambition, and fail to realize that they do not stand to gain as much as he does from the firm's success.

An old leader may feel increasingly vulnerable, and trust people less, as his physical powers decline and younger rivals queue up to replace him (Kets de Vries 1987). He is likely to demand increasingly stronger expressions of loyalty, which progressively fewer members are prepared to make. The loyal 'in group' begin to spy on the leader's behalf, creating a general atmosphere of distrust—particularly when people are unsure who is 'in' and who is 'out'. Indeed, the emergence of a secretive clique around the leader seems to be a hallmark of a monitoring system in the process of corruption.

The cultural environment of the firm is influenced mainly by the family and the nation-state. The family has a crucial role in manipulating cultural values early in the individual's life. Basic ideas about personal property, for example, are developed by understanding what is shared and what is not within the family unit. Perhaps most significantly, the religious affiliation of the parents will permeate attitudes to work and honesty—attitudes that will endure because of the loyalty factor noted earlier. A favourable family background can provide the firm with a steady supply of young recruits already imbued with many of the values it requires. Most especially, it can supply the firm with future leaders. A family culture that emphasizes trust in other people, the need for justice, and the importance of self-discipline in the exercise of power will generate leaders well equipped for moral manipulation.

In a stable democratic society the State is normally less influential than the family, but in an emerging nation with a one-party system the State is likely to be a potent influence. In some cases the State may even use its influence over the media and over the education system to discredit 'traditional' family culture. 'Modernization' may involve changing the vocabulary, and even the language, in which moral issues are presented. Modernization is typically geared to promoting an ethic of national unity and solidarity, undermining traditional regional or tribal animosities.

In a modernizing society control of State propaganda is to some extent 'up for grabs' between competing interest groups—business, trade unions, religious groups, and so on. The ability of business to win military or popular support, and to repel the challenge from labour interests, may be crucial in providing a favourable cultural environment under which reforms can be carried out.

It might be thought, from the preceding remarks, that any

family or national culture that was deliberately 'pro-business' would favour the moral values required by a firm. But this ignores the fact that what people believe is good for business is itself a cultural product. Even within 'pro-business' culture, there are two distinct sets of beliefs which reflect different views of human nature and, consequently, different management strategies. These beliefs pertain to the key issue addressed in this book—namely the possibility of engineering trust through moral manipulation as an alternative to monitoring.

The pessimistic view of human nature says that people can never really be trusted. Only the threat of punishment is really effective. Sometimes an all-seeing God who avenges unrepentant sinners with eternal damnation may be invoked to supplement material penalties, but this does not usually alter the basic approach. It may, however, allow the 'chosen few' to do business with each other on informal terms. But even then the few may be watching each other closely to check whether they really do behave as one of the chosen would do.

A more optimistic view of human nature sees business life as providing positive emotional rewards through interesting jobs and companionship at work. It emphasizes that placing trust in other people reduces the complexity of decisions and hence the level of stress, so that the quality of decisions is improved. This view too has a religious dimension, but one that emphasizes the values of community life rather than the importance of personal salvation.

Overall, therefore, moral manipulation within the firm is most likely to occur where there is a leader of appropriate age and experience, reared in a family with a strong commitment to high-trust religious values, and operating in a stable democratic State which is reasonably sympathetic to values of this kind. Conversely, heavy reliance on monitoring is most likely to be found in firms with very old or very young leaders who distrusted their parents or guardians and assimilated few moral values from them, and which operate in either an anti-business environment, or a pro-business environment in which a low-trust culture prevails.

13.7. Inter-Group Co-operation

So far it has been implicitly assumed that manipulation will be focused simply on the group which the leader controls. But there is a powerful reason for the leader to seek to influence the cultures of other groups as well. The gains from trade between groups are often greater than the gains from trade between individuals—differences in the resource endowments of the territories occupied, for example, often generate stronger comparative advantages than do personal differences in abilities or tastes. Trade is promoted by low transaction costs, and the minimization of transaction costs is a leadership function. If both the groups in a bilateral trade are subgroups of a larger group then the engineering of trust is the responsibility of the leader of the overall group. But if there is no encompassing group then the leader of the group with the most effective culture has a comparative advantage in engineering trust between them.

In some cases the leader of the superior group may work through the leader of the inferior group, converting him to his morality and relying on him to convert his followers. In other cases the followers may be addressed directly, with a view to isolating the leader and then deposing him. The latter strategy could, indeed, be so successful that the followers actually request to join the superior group, and so the inferior group loses its identity. Its territory and its resources are acquired by the superior group through entirely voluntary means.

Territorial expansion based on centralized leadership may generate diseconomies, however, because the superior leader has fewer advantages in dealing with an unfamiliar group with which it is difficult to maintain face-to-face communication. Arm's-length trade with an independent but friendly and trustworthy group may be more efficient than the extension of the group itself.

The expansion of a manipulative ethic through friendly relations with other groups has no parallel in the case of monitoring. The extension of sophisticated monitoring systems to other groups necessitates the monitoring of the monitors, since the monitors of the other groups cannot themselves be trusted. Just as monitoring within a group is essentially inquisitorial and coercive, so the expansion of monitoring requires the centralization of considerable material power within a hierarchy.

So far as the firm is concerned, this means that a manipulative ethic is well adapted to a strategy of quasi-integration in which friendly long-term relations are developed with major customers and suppliers (Blois 1972; Macaulay 1963). Monitoring strategies, however, tend to favour full forward and backward integration so that the hierarchy of supervison can be extended by ownership links. Without full ownership of the related activities their managers would not normally consent to detailed supervision.

Similar considerations apply where collusion is concerned. A manipulative ethic of solidarity promoted by one particular leader can be sufficiently influential to unify a cartel. But in the absence of a moral leader, monitoring mechanisms of sufficient strength cannot normally be established without horizontal integration of the members through merger and acquisition.

When viewed from the standpoint of the economy as a whole, the infusion of manipulative ethics into corporate culture is likely to generate wide-ranging long-term co-operative arrangements between small and medium-sized firms, with a number of key firms, led by a charismatic and committed manager, acting as the hubs of local networks. The diffusion of such an ethic into an economy which has hitherto been deficient in moral leadership will lead to the radical reorganization of indigenous firms whose leaders embrace the ethic, and acute competitive problems for some of the highly integrated firms whose leaders resist it. It should not, of course, be assumed that manipulation is universally superior to monitoring, and there may well be cases where integrated firms successfully resist the challenge by manipulative rivals. The scenario suggested here does, however, have striking similarities with the recent experience of Japanese firms investing in the USA, in which the manipulative strategies of the Japanese outperform the more rigid monitoring strategies of indigenous firms. Although a multiplicity of factors are at work, the difference between manipulative and monitoring strategies seems to be of crucial significance for recent developments in management styles (Casson 1990*b*).

13.8. Competition between Cultures

A successful culture can constitute an important source of national competitive advantage (Casson 1990*b*; Porter 1990). When the national culture and indigenous business cultures are symbiotic, production will be internationally competitive, as reflected in conventional macroeconomic indicators such as per capita Gross

Domestic Product, a strong exchange rate, low inflation rate, and so on.

It has been argued in this book that a high-trust culture is generally superior to a low-trust one. The theory shows, however, that there are certain circumstances in which engineering a high-trust culture is actually more expensive than operating a monitoring system. A truly successful culture must contain sufficient diversity to recognize this, and accept the need for monitoring in certain cases. It is unreasonable for people to expect to be trusted by everyone else all the time.

Cultural diversity provides the ability to adapt to changing conditions. A country with a rich and varied tradition can respond to new situations by resurrecting old concepts and ideas used by previous generations. A country with a more limited cultural base, however, must import ideas from elsewhere (see below).

Diversity must be achieved within an underlying unity, however. If advocates of different ideas cannot tolerate their opponents then the society will degenerate into factions. Culture change will become linked to major shifts in the balance of economic power, and conflicts between rival vested interests will create inflexibility. This will be reflected in, amongst other things, a slow pace of structural adjustment within the economy.

A country with a successful culture can export it to other countries. This is likely to be a slow process, however, because in the host country it will be the younger generation rather than the older one which is most susceptible to this influence. Nevertheless, in the long run the images of life-style used in the successful country's consumer advertising, the management practices of its multinational enterprises, and the behaviour of its emigrants (especially skilled workers and professionals) will all influence overseas economies (Earl 1986).

A host economy which is not performing particularly well, and where there is an ambition to 'catch up' with the most successful economies, is particularly vulnerable to influence of this kind. Citizens of the host economy may voluntarily submit to foreign ideological influence in key economic areas (especially if this influence is mediated by their own leaders, or by an indigenous 'modernizing' business élite). Less successful cultures may therefore be modified by the impact of more successful ones over time.

If the host-country culture is relatively weak, then in the long run it may be completely annihilated by foreign influence. If it is basically strong, it is likely to survive, but in a modified form. It may even have the power to influence the source-country culture, through the feedback of the attitudes and experiences of source-country exports, investors, expatriates, and emigrants. A successful culture is therefore also liable to undergo change. It will become dissociated from its specific geographical and ethnic roots and become more of an international culture, synthesizing the traditions of the source country with those of the principal host countries to which it has been transferred.

Competition between rival cultures could, in principle, drive the least efficient cultures to extinction. Societies which remain committed to inefficient cultures would face extinction by economic or military means. Those which survived would do so only by abandoning their own traditions and absorbing more efficient cultures from abroad. Cultures are, however, complex entities and there is no reason to believe that the process of selection through cultural competition, if it exists, has yet fully worked itself out. This means that the most successful cultures may still contain inefficient elements. They succeed in spite of some of their attributes rather than because of them. This point has important implications for the analysis of recent cultural changes affecting economic policy, which are taken up in the following chapter.

13.9. Summary

This chapter has examined the relation between national culture and business culture. Firms can to some extent 'free- ride' on national culture by picking up and emphasizing general moral principles such as dedication, loyalty, and integrity. Provided chief executives do not distort these values too much in applying them to their own ends, their manipulative efforts can assist the national leader—good employees make good citizens too. Firms do, however, compete for allegiance with other subnational groups, such as the family and the local community, and can potentially damage these groups by reducing people's commitment to them.

The success of business activities can encourage other countries to imitate national cultural traits that are believed to hold the keys to economic success. In the 1960s and 1970s US culture was widely imitated by countries wishing to 'catch up' in technological developments. In the 1980s Western countries have begun to imitate Japan, in the hope of raising productivity through better industrial relations. As the economic environment changes so cultural traits go in and out of fashion, it seems.

Imitation is to some extent a sign of weakness, because it reflects the lack of appropriate traditions in the native culture. For this reason, some countries are reluctant to imitate anything. If their own cultural base is very narrow, however, then unwillingness to learn from others because of misplaced pride may spell economic disaster.

The only countries strong enough to survive in a volatile environment without recourse to imitation are those with built-in flexibility provided by cultural diversity. Diversity does not signify a proliferation of unrelated ideas promoted by different factions, but rather the preservation of different traditions through education, political debate, and social ritual. These traditions need to be integrated within a unifying view of national heritage.

In recent years the cultural base of Western democracies seems to have narrowed quite significantly, due to a mistaken belief that certain traditions are incompatible with technology-driven economic growth. This may explain the apparent paradox of advanced industrialized economies having to learn from a traditional society like Japan—a society which is, moreover, a relative late-comer in terms of technology. Patterns of cultural change are continually evolving, however, and the 1990s provide an opportunity for Western nations to resurrect elements of their own cultural legacy. This may be a more effective way of engineering cultural change in the West than continuing to import ideas from the East and hoping that they will somehow adapt to Western conditions.

14

Policy Implications

14.1. Introduction

This chapter examines the implications of cultural engineering for national economic policy. It is argued that the supply-side economic thinking which has dominated Western government policies in the 1980s has had some distinctly adverse consequences for national economic performance. This is not to deny that many of the changes associated with the supply-side policy revolution were long overdue. There is a problem, however, in that supply-side policies have tended to undermine trust between economic agents and so raise rather than lower transaction costs. Competition between incentive-driven selfish players is not sufficient to guarantee high performance.

The problem seems to be that many governments have imported supply-side policies as a 'package' from the USA (and, in some cases perhaps, from the UK too). The package comes laden with certain cultural presuppositions—amongst them the view that because people cannot be trusted, formal methods of enforcing competition between them are essential in all areas of activity. The presupposition is taken from US culture, and may well be inappropriate elsewhere. The US emphasis on the law as an enforcement mechanism is a related cultural attitude which may not 'travel' well.

There is also a certain irony in the way that Western governments have borrowed their ideas on economic management from the USA at a time when many of the most successful private Western enterprises are borrowing their ideas on internal management from Japan. As a result, private enterprise has, to some extent, become more 'human' and informal, at the same time that the public-sector management has become less so.

Because Japanese culture is, in some ways, more obviously idiosyncratic than US culture, the dangers of copying Japanese attitudes without adaptation are very evident. The same is not true of the USA, however, because the effects of the US immigrant 'melting pot' of the nineteenth and early twentieth centuries has been to create a more cosmopolitan culture there. Nevertheless, the culture created by European immigrants in the USA is very different from the cultural legacy of Europe itself which those immigrants left behind. Transferring US attitudes on economic management to Europe needs the same kind of adaptation as does the transfer of Japanese attitudes if it is to be done successfully. The uncritical imitation of US economic policies which occurred in Europe during the last decade can hopefully be replaced by something more appropriate instead.

14.2. Supply-Side Policies and Decentralization

The supply-side revolution began in the USA in the 1970s, although the theoretical ideas that underpin it go back many years. It was taken up by Mrs Thatcher's Government in the UK in 1979 and since then has influenced other European governments in varying degrees. It is not confined to Western industrialized countries—many developing countries have switched to more liberal trade policies, often under the influence of World Bank structural adjustment programmes. Even countries such as China, where political change has been erratic, have followed a fairly consistent policy of economic liberalization during the 1980s.

Ostensibly supply-side economics is inspired by a view of human nature which emphasizes the intelligence and moral autonomy of the individual. Intelligence is not confined to an élite on this view. If it were, decision-making would have to be centralized in the hands of the few people capable of making sensible decisions. On the contrary, it is claimed, intelligence is widely dispersed. It is unnecessary for all information collected by individuals to be transmitted to a central authority before action can be taken. Instead, individuals can act immediately on the basis of their own information. Such decentralization of decision-making does not result in anarchy so long as market prices are available to guide individuals about the relative scarcities of resources within the economy.

In conventional economics information on relative scarcity is provided by prices set by Walrasian auctioneer. This is a rather contrived view, however, since no real negotiations take place between individual traders and no trade at all takes place until prices have been adjusted to equilibrium. Price-setting becomes a kind of game which could just as easily be carried out by planners within a centralized system.

A more realistic account of the market process is provided by the Austrian school, and it is this account that is typically favoured by enthusiasts of supply-side economics. The Austrian view emphasizes that the provision of price information can itself be decentralized: price quotations are provided, not by a Walrasian auctioneer, but by entrepreneurs engaged in speculation and arbitrage (Kirzner 1973).

The supply-side view of Western countries' economic performance during the 1970s is that economic growth slowed down because market forces were increasingly inhibited. This impeded adjustment to real shocks—such as the oil price rise and the growing competitiveness of Japanese industry. The adjustment of domestic relative prices to reflect the new international realities was inhibited by excessive regulation and the imposition of distortionary taxes and subsidies. State industries producing non-tradable outputs, in particular, were doubly protected—by the absence of import competition, and by the availability of State funds to cover operating losses. While nothing much could be done about the first form of protection, which resulted mainly from the nature of the product, something could be done about the second, which resulted from the nature of ownership. The obvious policy response was to roll back the frontiers of the State by deregulating markets and privatizing State-run activities.

There is, however, a paradox in the way that supply-side policies have been implemented. For in many cases radical 'liberalization' has been accompanied by growing centralization of decision-making and increased power of the State. In the UK, for example, industrial relations are now more heavily regulated than at any previous time in recent history. Central government has not only gained tighter control over local authority spending, but has dictated highly specific and detailed policies to health and education authorities, deliberately weakening intermediating institutions of a quasi-autonomous nature.

The preceding analysis suggests a simple resolution of this paradox.

Political advocates of supply-side policies do not derive their ideology from the assumed intelligence of human beings as much as from their assumed selfishness. Intellectual arguments about the decentralization of decision-making are employed because they are useful for rhetorical purposes in supporting the pro-market position. But the real reason why market forces are desired is that they help to discipline people who cannot otherwise be trusted.

Selfish people always face conflicting objectives in a world of scarce resources. This conflict creates a strong incentive to cheat, and in the absence of punishment it is assumed that they will do so. The fact that people may be both selfish *and* intelligent only makes the problem of distrust worse. Selfishness with guile is more to be feared than selfishness alone.

14.3. Economic Freedom in a Competitive Economy

Competition is seen by supply-side advocates as a device which can resolve this problem. Markets are desired because they provide an institutional environment in which competition is most effective. Competition prevents sellers from demanding too high a price. It allows buyers to adopt a 'divide and rule' policy by playing off one seller against another. Similarly it allows sellers to play off buyers against each other in order to prevent the price from falling too low.

In naïve statements of economic liberalism competition is claimed to be entirely sufficient as a constraint on individual behaviour. Conscience is redundant because competition will reconcile individual greed with the wider social interest. Indeed, by liberating greed from conscience, it is claimed, people can be energized by materialistic motives to work harder and to seek out new markets. As a result, the economic system works better without moral self-restraint than it does with it. The magic of the market's 'invisible hand' effects a moral transformation in which the 'private vice' of greed becomes the 'public virtue' of enterprise and innovation.

The promotion of competitive markets not only liberates people from conscience, it is claimed, but provides them with another kind of freedom too. This is the freedom to participate in a wide range of transactions without having to seek permission first. From the consumer's point of view, the wide range of choice encourages the development of discriminating tastes. From the producer's point of view it means that the production of key commodities is no longer reserved, through statutory monopoly, for the privileged few. There is freedom of entry into each industry which allows each producer to specialize according to his comparative advantage.

Economic freedoms of this kind have played a prominent role in the political rhetoric of the 1980s. Their relevance to the citizens of centrally planned economies, coping with the insane rigidities of overcentralized economic systems, cannot be doubted. But while economic freedom may be a *necessary* condition for a reasonable quality of life, it is more questionable whether it is a *sufficient* condition too.

From a producer's point of view, freedom of entry provides only limited opportunities for profit if it merely provides the freedom to compete with powerful established firms. To achieve really significant material gains it is normally necessary to enter an entirely new industry, or to create a new product. It is the freedom to establish a new monopoly that is really prized in material terms. This attitude is well articulated in the competitive-strategy literature (Porter 1990), where competititon is interpreted mainly as competititon to secure monopolistic positions. Moreover, since monopolistic advantages are continually under threat of obsolescence from imitators, or from subsequent innovations, monopoly rents provide little security in the long run.

Similarly, where consumers are concerned, greater consumption is often desired in order to differentiate the customer from others—to establish higher social rank through the conspicuous consumption of exclusive fashion goods. Thus abundant supplies of cheap goods do not afford the anticipated benefits to individual consumers because of the disappointment they experience when they realize that everyone else can just as easily afford them too.

Perhaps the most fundamental problem with over-emphasizing economic freedom, though, is that it plays down the emotional dimension of life. By emphasizing the material at the expense of the emotional, producers are denied feelings of satisfaction stemming from the intrinsic moral worth of their product. Similarly consumers are denied any context in which their acts of consumption (or self-denial) carry any higher moral significance. The cost of liberating people from conscience is that they become morally desensitized. From a utilitarian standpoint, they become worse off because the pleasures of emotional life are denied them.

14.4. The Influence of District on Leadership Style

A leader who cannot bring himself to trust other people perceives the cost of moral manipulation as very high. He is therefore strongly disposed to monitoring. He will favour an organization in which information is centralized for purposes of evaluating other people's performance. Not only does he not trust other people in general, but he cannot even trust a few people to do his monitoring for him. The leader will also favour strong interpersonal competiton, sustained by making people rivals for one another's jobs. The intensity of competition may be further raised by creating a pool of unemployed or marginal members who are ready to step into other people's jobs at short notice. Competition helps the leader to get 'value for money' from his followers, along the lines indicated earlier. It also provides tougher sanctions against people who are found to have underperformed. By increasing the versatility of his followers, the cost of replacing a poor performer is reduced because there is a large number of replacement candidates to choose from. By reducing the cost of replacement, the credibility of threats against poor performers is increased, and so potential slackers will be more worried about the consequences of detection. This means that the costs of monitoring can be reduced because a lower probability of detection is required to sustain hard work.

A leader who employs rigorous monitoring strategies backed by severe sanctions is unlikely to be popular. He may fear collusion aimed at deposing him. Competition can provide valuable reassurance in this respect. Collusion is often organized informally and in secret. These informal arrangements are easiest to sustain in small, stable, and compact groups (as indicated earlier). The leader will therefore try to break these informal bonds by setting members of these groups against each other. He will use competititon to destroy the informal infrastructure by which small groups operate.

Unfortunately for overall efficiency, however, these informal bonds also underpin the everyday co-ordination of activity within the group. The leader's strategy not only undermines collusion, therefore, but also the productivity of the group as a whole.

Sometimes the strength of these informal mechanisms may be sufficient for them to resist the leader's attempt to break them. In this case the leader can employ inter-group competition instead. He can lead the different groups to behave as rival factions by stirring up jealousies between them. Having factionalized the economy, all he needs to do is to ensure that his own faction is very much stronger than any of the others. He does this by securing for his faction a monopoly of certain key activities—such as law enforcement—on which all the other groups depend.

A leader who basically trusts his followers will, by contrast, rely much less on monitoring and much more on manipulaton. Because he has less need for monitoring, he has less need for hierarchical organization. He also has less need for competition because he believes other people are more likely to settle spontaneously on fair prices and to honour the contracts they make. There is less need, therefore, to play off rivals in negotiation and to threaten to replace people who are found to have cheated. A leader who trusts other people is also likely to behave in a more trustworthy way himself, both because it pays to reciprocate trust out of enlightened self-interest and because he believes his own high-trust rhetoric. As such, he has less to fear from collusion, both because his rhetoric deters collusion, and because there is nothing unjust in what he does to encourage opposition anyway. The high-trust leader can maintain discipline over potential factions by his own moral authority. He can reinforce this with an ethic of unity, which encourges the leaders of potential factions to co-operate with each other as a cohesive élite—rather than fight with each other.

A leader who trusts other people can also tackle an issue which a leader who does not trust other people cannot even begin to address satisfactorily. This is the problem caused by selfish and dishonest behaviour in situations where contracts cannot readily be negotiated and enforced. Petty crimes such as thefts, intimidation, and traffic offences are difficult to contain through monitoring mechanisms because of the enormous costs of detection—in particular in finding reliable witnesses to identify the culprit. Moral manipulation by a leader who believes that people can be trusted solves this problems directly by encouraging people to police their own actions.

14.5. UK Economic Policy in the 1980s: A Case-Study in the Effects of Distrust

The strategies employed by the Thatcher Government in the UK provide an admirable illustration of some of the general points made above. Although the official line has been that the Thatcher reforms involve supply-side policies geared to strengthening material incentives and allowing competition greater play, there have been discernible trends that point in a rather different direction.

Much of recent UK policy is a reaction to the failed economic consensus of the 1960s. Towards the end of the 1960s the social contract between the trade unions, the employers, and the government broke down in the UK. In the early 1960s responsibility for economic management was shared by three main respresentative organizations—the Trades Union Congress, the Confederation of British Industry, and the government. Short-term economic management relied heavily on prices and incomes policies, whilst long-term policies were based on economic planning on the French model. Incomes policies failed, however, to reconcile workers' rising real-wage aspirations with slow productivity growth. Incomes policies were viewed by trade unions purely as instruments of wage restraint, and whenever incomes policy slacknened off inflation acclerated to restore prices to what they would have been had the policy not been in force. Long-term planning failed to achieve aggregate productivity growth comparable to that in France because of structural rigidities—for example, trade-union pressure prevented the shedding of labour in obsolescing industries. The underlying distrust between employers' and workers' representatives continued unabated.

The Thatcher Government chose not to address head-on this legacy of mutual distrust. Rather it displayed strong distrust of its own towards at least one of the groups involved. It opted to assert the dominance of government over the other factions. Instead of eliminating factionalism, government sought to strengthen its position by exploiting more fully its legislative powers. More specifically, it sought to impose greater checks on trade unions by removing legal immunities and outlawing secondary strike action. The measures reflected a narrowly materialistic view of trade-union bahaviour: namely that trades unions exploit their immunities to strengthen their exercise of labour monopoly power.

After 1979 prices and incomes policies were initially replaced by a policy of monetary restraint, and trade-union militancy was further inhibited by high unemployment. More recently, though, wage restraint has been reintroduced indirectly in the public-sector.

The government has begun to regard professional associations involving public-sector employees as engaged in collusive activities akin to those of trade unions. It has chosen to ignore the role of these associations in training and quality control and to draw attention to their 'restrictive practices' instead. The professional ethos of service to the community which has traditionally inspired loyalty and commitment is seen purely as a device for fostering collusion and as something which must therefore be undermined. Whilst established professionals have usually maintained their loyalty, the reduction in social status, coupled with worsening relative pay, has discourged new entry, and built up even more serious potential problems for the future.

Government distrust of public employees is nowhere more apparent than in the proliferation of formal monitoring systems within the public sector. In many cases the government has intervened directly in the affairs of hitherto autonomous public bodies to ensure that these systems are introduced. Large investments have been made in computerized accounting systems to measure performance more objectively within the public sector. The introduction of these expensive systems has been presented as a necessary measure to bring the public sector into line with the more efficient private sector. But what has not been mentioned is that in the most sophisticated private firms accounting systems are complemented by 'human relations' management designed to deal with aspects of performance which are difficult to quantify. Exclusive emphasis on formal monitoring creates obvious biases in favour of short-term quantifiable results. The costs of long-term decline in underlying quality, due to poor morale, for example, may remain hidden in the short run, but they cannot be disguised for ever.

The government's preoccupation with monitoring, and its neglect of morale effects, is part of a still broader phenomenon, namely its emphasis on formal procedures and objective measures at the expense of informal mechanisms and subjective assessments. This is reflected in its determination to legislate for trade-union reform, and to institute further bureaucratic monitoring within an already bureaucratic public sector. One interesting manifestation of the preference for formality is the enormous amount of legislation that the government has brought before parliament. In the short run, of course, this has the desirable effect, from the government's point of view, of stifling debate on specific issues, although it has the unfortunate long-term effect that hasty legislation only builds up problems for the future.

Another attitude which is particularly apparent in the government's approach to public-sector issues is that only behaviour motivated by material self-interest is believed to be efficient. Profit-seeking seems to be synonymous with efficiency, and altruism with inefficiency, so far as the government is concerned. These attitudes tie in with the remarks about competition made earlier. If competitive constraints can substitute for self-control then material greed, when liberated, can create a powerful drive for efficiency. The converse would seem to be that when greed is inhibited then the ruthless pursuit of efficiency is ruled out altogether. If this very crude psychology is rejected, however, then it becomes obvious that a strong commitment to altruism may be just as powerful in promoting efficiency directed to altruistic ends as is greed in promoting efficiency directed to selfish ends. It is simply that the efficiency criterion used by the altruist will normally attach greater weight to the social costs involved in the single-minded pursuit of any one narrowly defined goal.

14.6. National Missions of Unity

The preceding interpretation of UK economic policy may well be regarded as contentious, but even those who basically agree may still ask what could have reasonably been done instead. Is there any politically feasible mechanism for improving UK performance by engineering greater trust within the economy?

To unify factions such as labour and management, and to encompass all of the factions represented by different professional groups, a fairly powerful sense of mission is likely to be required. The mission may be related either to general moral principles of universal relevance, or to more parochial concerns, say of a nationalistic nature. Nationalism has the advantage that it relates directly to the political life of a country, but has the disadvantage that it may simply create a larger faction—one nation— at odds with the rest of the world. Since the preceding discussion has focused on the UK, the question of a national mission for the UK is what is addressed here.

Historically many nations have built up unity around the perceived need to catch up with technologically advanced economies, but because of its pioneering role in the Industrial Revolution the UK has never had a very strong ethos of this kind. Part of the Thatcher Government's 'realism' was that it recognized how far the UK has recently fallen behind its technological competitors, but it cannot be claimed that in the UK technological missions generally have enjoyed much success. The prestige-oriented research into nuclear energy and aerospace in the 1960s was certainly not a commercial success.

Indeed, it could be argued that the anti-technological missions of conserving the landscape, preserving monuments, and embellishing social traditions have a far stronger grip on popular imagination in the UK. In the technology-driven 1950s and 1960s these concerns seemed quite out of place in the modern world, but the recent world-wide revitalization of 'green' issues has changed this. Greater emphasis on sustainable environments may afford the UK an opportunity to take an international leadership role again. An initiative of this kind would, however, call for a dramatic reversal of the trend in government policy over the last decade. It would damage vested interests in pollution-intensive industries, and so would be unlikely to command universal support. But it would have the great advantage, nevertheless, of resonating with some of the traditional concerns of UK society.

Another traditional concern in the UK has been the need to reconcile industrialization with social welfare. Late Victorian enthusiasm for Practical Christian Ethics (even amongst non-believers) inspired many impressive philanthropic initiatives in the field of education, health, temperance, housing reform, protection of children, and care for the destitute.

Social missions of this kind are not without their costs in terms of private consumption, of course. This becomes very clear when the problems created by the post-war engineering of the Welfare State are addressed. Escalating public expenditure led to high marginal tax rates, and escalating public employment in services crowded out manufacturing and discouraged exports. The root of this problem does not seem to have been the nature of the mission itself, however—for example, popular support for the principles of the National Health Service has always been strong despite the rising costs that modern methods of treatment involve. Rather it appears that reliance on the formal bureaucratic methods of the State to implement the welfare agenda may have been at fault. One aspect of this is that legislation created vested-interest groups of claimants who sought to lobby for greater material benefits for themselves. Tax-payers, as donors, had little sympathy for many of these demands, partly because the coercive nature of taxation created resentment, and partly because salaried professionals increasingly stood between the donor and the recipient, replacing a personal relationship by a much more impersonal one.

There also appear to have been deficiencies in professional management in public services. Some of these may have been attributable to the lack of material incentives, as the Thatcher Government has alleged, although many of the professionals involved in public services are probably less sensitive to purely material incentives than most. In any case, some of the deficiencies of public management have been mirrored in deficiencies of private management too. Managerial inefficiency in the UK is not a problem unique to the public sector.

It may be, therefore, that the mission of creating a Welfare Society is basically sound, provided that the managerial problems involved are more seriously addressed. Because there are genuine problems of relying exclusively of public rather than private provision, however, it may better to think in general terms of a Welfare *Society* rather than specifically on a Welfare *State*. The fact that a Welfare Society may also be more environmentally sustainable than a Consumer Society, engineered on Thatcherite lines, underlines the case for considering a mission of this kind to a viable alternative to the policies pursued in the UK over the last ten years.

15

Summary and Conclusions

15.1. The Advantages of Moral Leadership

This book has presented a simple account of the ways in which culture can affect economic performance. It has developed a family of simple models which yield hypotheses of real explanatory power.

The basic idea is to distinguish between material and emotional rewards. Emotional rewards can be influenced by moral considerations, but material rewards cannot.

Morality, though ultimately a matter of individual conscience, has an important social dimension. The models in this book postulate that each individual is affiliated to a social group. Within a group, the moral outlook represents value-laden information which has the property of a public good.

There is a division of labour between leader and follower within each group. Group performance depends on the efficiency of co-ordination, for which the leader carries overall responsibility. Formal co-ordination involves explicit recorded agreements, such as contracts negotiated in a market. The contracts may specify, where appropriate, that the individual will work within a hierarchy and respond to the leader's command—this is a common arrangement, for example, within a team.

Both markets and hierarchies incur transaction costs because people cannot trust each other. Problems of mistrust can to some extent be solved by third parties. A classic example of this is the role of the head of State. The State holds hostage the freedom of individual citizens—it can imprison them at will. This allows the State to apply sanctions against cheats—provided they can be caught. The ease of detection and capture can be increased further if civil liberties—such as the rights to privacy and freedom of movement—are restricted.

This does not, ultimately, solve the problem, however, for it is still necessary to consider whether the head of State can be trusted. While it reduces the need to trust ordinary citizens, it increases the need of every citizen to trust the head of State. Unless the head of State is himself morally committed to self-restraint, he cannot be trusted not to abuse his power.

A leader of any kind, of course, needs to be trusted, but the trust that needs to be reposed in a purely moral leader who demands no hostages is much less—so long as his followers retain sufficiently independent critical faculties to reject spurious rhetoric. The leader legitimates by argument and by example the morality which he urges the followers to adopt. Trust engineered in this way is more compatible with followers' freedom than is the monitoring and hostage system.

15.2. A Summary of Manipulation Strategies

In a Prisoner's Dilemma it always pays to cheat whether your partner has cheated or not. If a leader can engineer a shared emotional commitment to honesty among his followers, however, then when the emotional and material rewards are combined, it may pay to be honest instead. Cheating becomes associated with guilt, and so the desire to avoid guilt motivates honesty. The individual concerned becomes effectively self-policing. Instead of relying on an expensive legal system, the potential offender becomes detective, prosecutor, and judge with respect to himself.

In some cases the emotional penalty of guilt may be sufficiently strong that it pays to be honest whatever your partner does. In other cases it only pays to be honest if you think your partner is going to be honest too—if you think he is going to cheat then it may still pay to cheat as well. This indicates that it is not only the emotional pay-offs that matter, but the perceived probability that the partner will be honest too. Under these conditions, announcements concerning the integrity of others are sometimes sufficient to produce a self-validating outcome, so that moral manipulation becomes superfluous.

In the absence of specific information about a person's previous behaviour, perceptions are likely to be based on the average behaviour of the population of people from whom the partner is drawn. In a one-off encounter that has no precedent, the estimate of the population probability must be a purely subjective one. As a result, there may well be a dispersion of perceptions of the population probability. But where the encounter is of a familiar type, for which many precedents exist, then individual perceptions are likely to converge on the population average. This average, however, reflects behaviour at a time when earlier beliefs were influential. These early beliefs, the models show, can become locked into subsequent behaviour—not in spite of, but because of, subsequent learning. If initial beliefs are sufficiently pessimistic, then people will be induced to cheat at the outset, establishing a record of cheating which suggests that honesty is never worthwhile. Initial pessimism allows the population to become entrenched in a cheat–cheat equilibrium in which all contact is eventually eliminated. By contrast, initial optimism can encourage honesty from an early stage, building up a record which encourages even greater honesty in future. This leads to an equilibrium of mutual honesty. The formal models show that there is a critical level of initial optimism which is necessary to sustain the honest equilibrium. Intelligent leadership may be able to engineer initial optimism and so ensure that it is the honest equilibrium that is attained.

15.3. Multi-Stage Decisions

An encounter between two members of a group is frequently preceded by, and followed by, other strategic decisions. The preceding decision is whether to venture out and go into an encounter, or whether to stay in to avoid it. (For continuing encounters the corresponding decision is whether to remain attached or to separate—i.e. to quit.) The following decision is whether to seek out a second encounter purely to 'settle a score' arising from the first.

The key point about the first stage is that the desirability, or otherwise, of avoiding an encounter depends critically on the probability with which the partner is expected to cheat. A high probability of cheating can discourage people from engaging in encounters altogether. This point is very important because the people who are most likely to suffer from cheating are those who always choose to be honest because their moral commitment is strong. The participation decision therefore tends to select against the potentially honest when the perceived probability of cheating is relatively high. This intensifies the mechanism, noted earlier, by which initial beliefs become entrenched in the equilibruim outcome. Initial pessimism becomes strongly self-validating because it discourages the morally committed from participating. Recorded outcomes are dominated by the uncommitted who are prone to cheat. Conversely, initial optimism encourages the committed to participate and so bolsters the proportion of honest encounters that can be recorded.

The important thing to be said about the final stage is that, in the absence of emotion, it is rarely optimal to pursue a cheat. The material costs of wreaking vengeance usually exceed any material rewards obtained. This in turn increases the viability of cheating, of course, and makes a cheat–cheat equilibrium more likely. The emotion of anger, however, operates in the opposite direction. It provides an emotional pay-off to vengeance which can outweigh the net material loss involved. Anger, when suitably directed, can therefore make the honest equilibrium more likely.

Anger can be used not only against other followers. It is particularly potent against the leader of a group. Emotional bonds between leaders and followers can frequently be intense—though usually only in one direction. The intensity of affection for and dependence on the leader makes the leader's betrayal of trust a very serious offence. In a compact group it is easy for followers to recognize that their anger is shared by others. Although a cool calculation of the expected material benefits of insurrection may indicate only a loss, the expected emotional pay-off may be high. In this case the leader may be threatened with action which, even if unsuccessful, may be extremely expensive in terms of his own material interests. In this sense, popular anger constitutes an important sanction which may prevent even an uncommitted leader from behaving unjustly towards his followers.

When information flows within a group are highly centralized, the leader may acquire a virtual monopoly of information about what goes on outside the group. This allows him to misrepresent the environment of the group, if he wishes, and to whip up hostility towards other groups. An inefficient or unjust leader may thereby transfer anger that might be directed against himself to external targets instead. By maintaining a degree of solidarity within his own group, he may actually improve their material well-being.

He can harness the solidarity to promote collusion which redistributes income from other groups.

The tendency for inefficient leaders to redirect anger is one of the main reasons why leadership is regarded with such suspicion by many people. If two leaders play this game against each other, group rivalry can escalate into unrestrained conflict. The answer, of course, is to establish internal democratic mechanisms by which information can be decentralized and bad leaders replaced. The abuse of leadership in centralized groups should not obscure the fact that effective leadership can promote both efficient performance and a just internal distribution of income.

15.4. Optimal Manipulation Strategy

The co-ordination of activities raises incentive problems of various kinds. In resolving any given problem the leader of a group has a choice between establishing a system of monitoring, which normally requires the use of hostages, and effecting moral manipulation instead. This choice is governed by comparative transaction costs, which in turn depend on the personality of the leader, the nature of the situation and the cultural environment within which the leader operates.

Because this book focuses on manipulation, monitoring strategies have been analysed in a relatively superficial way. Thus it has been assumed that no errors of observation occur and that, because people are risk-neutral, the size of hostage can be set sufficiently high to eliminate all cheating. Monitoring does, however, incur a fixed cost which is related to the size and disipersion of the group being monitored. It also depends on whether the situation affords natural hostages or requires hostages that have to be especially supplied.

So far as manipulation is concerned, there are diminishing returns to the intensity of manipulation because different people have different sensitivities. This means that as the intensity of manipulation increases, fewer additional people cross the threshold to honesty. Some residual cheating always remains and this creates an indirect transaction cost which has no parallel, under present assumptions, in the monitoring case.

Like monitoring, manipulation incurs fixed costs but, unlike monitoring, these depend very much on whether communication is face-to-face. Manipulation is also much more sensitive to the personal qualities of the leader—as determined by his family background and religious affiliation—and to the cultural environment in which he operates. This environment is determined by the values promulgated by the leaders of the other types of group to which his members belong. It is particularly important that when all the leaders share a commitment to high-trust culture then beneficial externalities reduce the costs of manipulation to each of those involved.

Efficient manipulation requires that the moral message is tailored to the nature of the incentive problem involved. An ethic of achievement and dedication is useful in reducing slacking, whilst an ethic of integrity reduces cheating on contracts and so helps to promote trade. The ethic of solidarity is more ambiguous, because solidarity is easiest to achieve in small groups. To maintain the viability of the large groups required to exploit economies of scale and a sophisticated division of labour, this ethic must be accompanied by an ethic of unity, for otherwise disintegration may occur. Solidarity is also ambiguous because it can easily be harnessed to promote unrealistic demands for distributive justice—demands often fuelled by a holistic moral view in which the legitimacy of many private property rights is questioned. A successful leader needs to exercise moderation and self-restraint in order to neutralize claims of injustice, and to make his followers aware of the potential economic penalties of repudiating totally the contractarian moral view in favour of a holistic approach.

15.5. Business Culture and the Internal Organization of the Firm

So far as the economic performance of advanced economies is concerned, the most crucial type of social grouping is the firm. Economists typically regard the firm as a monitoring mechanism, and devise predictions about its internal organization on this basis. When the firm is considered as a manipulation mechanism too, though, a rather different perspective on internal organization emerges.

It has already been noted that the dissemination of moral values depends crucially on face-to-face communication whereas monitoring can more easily be effected remotely. Moral argument also benefits from being customized to the target individual more than does monitoring, though this is of secondary importance. Manipulation requires fewer routine technical skills than does monitoring—fewer lawyers, accountants, and statisticians—and more articulate and charismatic people instead. It also favours small size and greater reliance on subcontracting to independent trusted firms.

On the whole manipulation promotes a more open and informal organizational structure, with employees being encouraged to engage in debate on fundamental issues rather than confine themselves to the technical details of their job. If the environment tailored to manipulation sounds more attractive to the employee, this reflects just another of the potential benefits of a manipulation strategy.

15.6. Prospects for Business Leadership.

Given the benefits of manipulation, it is difficult not to conclude that the cultural background of Western business leaders is sadly deficient. The societies in which they now operate discredited business leaders during the inter-war period, as a consequence of the international slump. The post-war legacy was popular support for trade-union collusion, and limited investment in business education.

US technological leadership encouraged imitation of US management methods. These methods reflect a strong belief that there is no real need to trust other people provided firms have access to a cheap and efficient legal system. The deficiency of this belief is apparent, however, in the way that vertical and horizontal integration has proliferated in the USA (since before the turn of the century) in order to provide internally more efficient monitoring and more effective hostages than were available from the legal system. The popular ethos of free entry by small firms conflicted with the barriers to entry created by the very conspicuous benefits of large-firm integration. This conflict between the ideology of free entry and the reality of large-scale integration is something which the supposedly efficient legal system, operating through anti-trust hearings, has never fully resolved.

European firms imported US management methods into an environment where the legal system was even less efficient, and where the legacy of the Protestant work ethic, on which US firms were able to free-ride, had been undermined by secular scepticism. Indeed, following two wars in which military leaders had been discredited by their incompetence and their disregard for the lives of their troops, there was major cultural crisis in Europe. This concerned the legitimacy of authority in general, and of the 'right to manage' in particular (the USA experienced a similar crisis during the Vietnam war, but its impact was less severe). The crisis of authority made monitoring strategies almost totally ineffective as managers were afraid to implement sanctions even when they had all the evidence they needed to do so.

The 1980s witnessed a period of deliberate cultural engineering as governments and firms struggled to overcome these problems. But whereas many firms, under the influence of 'management gurus' opted for a high-trust culture, most governments did the opposite. Trusting no one but the ideological extremist, some of them unleashed as much power as they could muster against the institutions they considered to be infected with the post-war cultural legacy. Conflict between management and unions, which bedevilled the 1960s and 1970s, was contained by high unemployment (though sometimes this was just the unintended effect of over-zealous monetary policy). But the underlying problem of mistrust in industrial relations was resolved in only a limited number of cases.

New conflicts have since arisen as a consequence of 'supply-side' policies, and there now seems little evidence of a unifying moral system of high-trust values emerging to contain them. The 'human relations' approach to management could in principle provide such a basis, but its extension to the family, the community, and the nation-state is problematic because of the very different types of encounter involved in these cases. Business education has, in any case, tended to move away from the human relations approach to leadership, and towards greater emphasis on adversarial strategies. With cultures so diffuse and fragmented, and with distrust a persistent theme, there is a danger that the 1990s could become a decade of despair, with no moral foundation by which even a basic code of business behaviour can be legitimated.

It is, of course, a prime responsibility of political leaders to address problems of this kind. But politicians can also benefit from the support of others—religious leaders, trade-union leaders, powerful businessmen, leaders of influential clubs and societies, and so on (Cochran 1985). Unfortunately, divisions amongst leaders in different spheres of activity (and amongst rival leaders in the same sphere too) have grown so large in some societies that the leaders have lost interest in promoting general moral values and now concentrate just on promoting factional interests instead. The supply of leaders meeting the broad criteria set out in this book is evidently very limited.

Indeed, popular perception of this shortage has probably contributed to contemporary Western scepticism about leadership in general. Hopefully, however, the academic study of business culture may be able to articulate the requirements of leadership in a systematic way. Through the integration of research and teaching, it may be possible to educate leaders for tomorrow who can do a better job than the leaders of today. Such education will, of course, need to emphasize the moral dimension as much as, if not more than, the technical one (Bloom 1987). It would be premature to place too much confidence in a programme of this kind, but the probability that it could be done provides some measure of hope for the future.

Notation

Greek—decision variables

For leader:

θ intensity of manipulation (Section 2.1)
ζ fine on slacking (Section 2.8)
φ prediction of crime rate (Section 4.2)

For follower:

μ strategy choice (Section 2.1); $\mu = 0$ (honesty), $\mu = 1$ (cheating), $\mu = 2$ (avoidance)

Roman—other variables (always in per capita terms where appropriate)

a (parameter) *symmetric co-ordination gain*—the material gain from mutual honesty compared to mutual cheating (Section 3.2)
b (parameter) *incentive to cheat*—the material gain to cheating (compared to honesty) when the other party is honest (Section 3.2)
c (variable) *manipulation cost* (Section 2.4)
c' (parameter) *monitoring cost* (Section 2.8)
d (parameter) *asymmetric co-ordination gain*—the net loss incurred by an honest person who is cheated, calculated after redistribution to the cheat has been taken into account (Section 3.2)
e (parameter) *cost of effort* (Section 2.1)
f (variable) *feeling of satisfaction* generated by participation in a mutually honest encounter (Section 6.3)
F (function) *distribution of sensitivity* across the group (Section 2.1)
g (variable) *guilt* (Section 2.1)
h (parameter) *material reward from mutual honesty* in an encounter (compared to the null reward from avoiding the encounter altogether) (Section 3.2)
i (parameter) *turnover of membership*—the probability that a representative member will leave the group between one period and the next (Section 12.3)
j (variable) *contribution rate*—the ratio of the number of members who contribute to the provision of a public (or common) good to the total size of group (Section 9.2)
k (parameter) *probability-sensitivity of guilt*—the probability-dependence of the critical level of guilt required to induce honesty (Section 3.2)

NOTATION

- m (variable) *size of team* (Section 7.3)
- n (parameter) *size of group* (Section 2.4)
- p (variable) *perceived probability* that a partner will cheat (Section 3.2)
- P (variable) *team-specific probability*—the perceived probability that a 'weak link' will occur in team production due to one or more of the other members cheating (Section 7.2)
- q (variable) *crime rate*—the average incidence of cheating within a group (Section 2.1)
- Q (variable) *cartel crime rate*—the average incidence of cheating by someone else within a cartel (Section 11.2)
- r (parameter) *discount rate* (Section 12.3)
- s (variable) *sensitivity* to manipulation (Section 2.1); *also* (parameter) *fair share*—the maximum share of output taxed by the leader that is compatible with vertical equity (Section 10.3)
- t (variable) *transaction cost* (Section 2.9)
- u (variable) *leader's utility* (Section 2.4)
- v (variable) *leader's objective* (Section 2.2)
- w (variable) *wage* paid by an intermediating team leader (Section 8.6)
- x (variable) *avoidance rate*—the proportion of group membership that chooses not to participate in encounters, or is disqualified by reputation effects from doing so (Section 5.2)
- y (parameter) *output* under dedication (Section 2.1)
- Δy (parameter) *marginal productivity of effort* (Section 2.1)
- z (variable) *number of contributors* to public good provision (apart from the representative individual) (Section 9.2)

Glossary

This glossary is for the benefit of readers who wish to study a single chapter out of context and would otherwise need to refer to earlier chapters. It defines some commonly recurring terms which are used in this book in a specific way.

Altruism: Concern by a leader with the welfare of his followers, in which followers are regarded as having interpersonally comparable utilities to which the leader gives equal weight.

Broad materialism: An altruistic leader's objective which includes followers' effort costs but excludes their feelings of guilt.

Cheating: Failure to conform with a moral norm articulated by a leader. The morality is a contractarian one.

Co-ordination: Pareto-improvement within a group.

Crime rate: Average incidence of cheating within a group—the ratio of the number of members who cheat to the number of members active in encounters.

Dedication: Hard work, i.e. honesty in the context of the supply of effort.

Follower: A member of a group who is manipulated by the leader.

Guilt: A self-inflicted emotional penalty associated with cheating, induced through moral manipulation.

Honesty: Conformity with a moral norm articulated by a leader. The morality is a contractarian one.

Leader: Manipulator of the followers in the group.

Loyalty: An ethic which associates guilt with quitting membership of a group.

Marginal productivity of effort: Excess of per capita output when everyone in the group is dedicated over per capita output when one or more members slack.

Materialism: See Broad materialism and Narrow materialism.

Narrow materialism: An altruistic leader's objective which excludes both followers' effort costs and their feelings of guilt.

Non-co-operative game: An encounter in which each party's rewards depends upon the partner's choice of strategy as well as his own, and where mutually beneficial contracts between the parties cannot be made.

Prisoners Dilemma: A non-co-operative game in which it pays each party to cheat if their partner is honest *and* if their partner cheats as well.

Sensitivity: Responsiveness of the individual's emotional rewards and penalties to manipulation by the leader. It varies between individuals and is measured on a scale from zero to unity.

Slacking: Low intensity of work i.e. cheating in the context of the supply of effort.

Stackelberg leader: An individual engaged in an encounter with people whose responses to his own actions he can correctly predict, and who uses those predicted responses to calculate his own best strategy.

Utilitarianism: An altruistic leader's objective which includes both followers' effort costs and their feelings of guilt.

Bibliography

Akerlof, G. A. (1970), 'The Market for Lemons: Quality Uncertainty and the Market Mechanism', *Quarterly Journal of Economics*, 84: 488–500.

—— (1980), 'A Theory of Social Custom, of Which Unemployment may be One Consequence', *Quarterly Journal of Economics*, 94: 719–75.

—— (1983), 'Loyalty Filters', *American Economic Review*, 73: 54–63.

—— (1989), 'The Economics of Illusion', *Economics and Politics*, 1: 1–15.

—— and Yellen, J. L. (1985), 'Can Small Deviations from Rationality Make Significant Differences to Economic Equilibria?', *American Economic Review*, 75: 708–20.

Alchian, A. A., and Demsetz, H. (1972), 'Production, Information Costs and Economic Organisation', *American Economic Review*, 62: 777–95.

Arrow, K. J. (1974), *The Limits of Organization*, New York: W. W. Norton.

Axelrod, R. (1984), *The Evolution of Cooperation*, New York: Basic Books.

Barrientos, A. (1988), 'Economic Man, Mathematics and the Rise of Neo-classical Economics', *Ealing Working Papers in Economics*, No. 5.

Barry, B. (1970), *Sociologists, Economists and Democracy*, Chicago: University of Chicago Press.

Bartlett, R. (1989), *Economics and Power: An Inquiry into Human Relations and Markets*, Cambridge: Cambridge University Press.

Baxter, J. L. (1988), *Social and Psychological Foundations of Economic Analysis*, Hemel Hempstead: Harvester Wheatsheaf.

Becker, G. S., and Stigler, G. J. (1977), 'De Gustibus non est Disputandum', *American Economic Review*, 67: 76–90.

Becker, L. C. (1986), *Reciprocity*, London: Routledge and Kegan Paul.

Blois, K. J. (1972), 'Vertical Quasi-integration', *Journal of Industrial Economics*, 20: 253–72.

Bloom, A. (1987), *The Closing of the American Mind*, London: Penguin.

Booth, A. L. (1985), 'The Free Rider Problem and a Social Custom Model of Trade Union Membership', *Quarterly Journal of Economics*, 99: 253–61.

Brubaker, R. (1984), *The Limits of Rationality: An Essay on the Social and Moral Thought of Max Weber*, London: Allen and Unwin.

Buchanan, J. M. (1965), 'An Economic Theory of Clubs', *Economica*, NS, 32: 1–14.

Campbell, T. D. (1971), *Adam Smith's Science of Morals*, London: Allen and Unwin.

Casson, M. C. (1982), *The Entrepreneur: An Economic Theory*, Oxford: Martin Robertson.

—— (1990*a*), 'Economic Man', in J. Creedy (ed.), *Foundations of Economic Thought*, Oxford: Blackwell, 3–27.

—— (1990*b*), *Enterprise and Competitiveness: A Systems View of International Business*, Oxford: Clarendon Press.

Chell, E. (1985), *Participation and Organization: A Social Psychological Approach*, New York: Schocken Books.

Church, R., Outram, Q., and Smith D. (1989), 'The Militancy of British Miners: Interdisciplinary Problems and Perspectives', *University of Leeds School of Business and Economic Studies*, Discussion Paper Series 89/6.

Coase, R. H. (1937), 'The Nature of the Firm', *Economica*, NS, 4: 386–405.

Cochran, T. C. (1985), *Challenges to American Values: Society, Business, and Religion*, New York: Oxford University Press.

Cohen, M. D., and Axelrod, R. (1984), 'Coping with Complexity: The Adaptive Value of Changing Utility', *American Economic Review*, 74: 30–42.

Collard, D. (1978), *Altruism and Economy*, Oxford: Martin Robertson.

Dahrendorf, R. (1973), *Homo Sociologicus*, London: Routledge and Kegan Paul.

Drakopoulos, S. (1988), 'Two Levels of Hedonistic Influence on Microeconomic Theory', *University of Stirling*, Discussion Papers in Economics, Finance, and Investment No. 150.

Earl, P. E. (1983), 'The Consumer in His/Her Social Setting: A Subjectivist View', in J. Wiseman (ed.), *Beyond Positive Economics?*, London: Macmillan, 176–91.

—— (1986), *Lifestyle Economies: Consumer Behaviour in a Turbulent World*, Brighton: Wheatsheaf.

Elster, J. (1979), *Ulysses and the Sirens: Studies in Rationality and Irrationality*, Cambridge: Cambridge University Press.

—— (1982), 'Sour Grapes: Utilitarianism and the Genesis of Wants', in A. Sen and B. Williams (eds.), *Utilitarianism and Beyond*, Cambridge: Cambridge University Press, 219–38.

Etzioni, A. (1988), *The Moral Dimension: Towards a New Economics*, New York: Free Press.

Felix, D. (1979), 'De Gustibus Disputandum Est: Changing Consumer Preferences in Economic Growth', *Explorations in Economic History*, 16: 260–96.

Freeman, R. B., and Medoff, J. L. (1984), *What do Unions Do?*, New York: Basic Books.

Friedman, J. W. (1977), *Oligopoly and the Theory of Games*, Amsterdam: North Holland.
—— (1986), *Game Theory with Applications to Economics*, New York: Oxford University Press.
Fudenberg, D., and Kreps, D. (1987), 'Reputation in the Simultaneous Play of Multiple Opponents', *Review of Economic Studies*, 54: 541–68.
Furnham, A., and Lewis, A. (1986), *The Economic Mind: The Social Psychology of Economic Behaviour*, Brighton: Wheatsheaf Books.
Gambetta, D. (ed.) (1988), *Trust: Making and Breaking Cooperative Relations*, Oxford: Blackwell.
Granovetter, M. (1985), 'Economic Action and Social Structure: The Problem of Embeddedness', *American Journal of Sociology*, 91: 481–510.
Halperin, R. H. (1988), *Economies across Cultures: Towards a Comparative Science of the Economy*, London: Macmillan.
Hamlin, A. P. (1986), *Ethics, Economics and the State*, Brighton: Wheatsheaf.
Hargreaves-Heap, S. (1989), *Rationality in Economics*, Oxford: Blackwell.
Harrington, J. E., Jr. (1989), 'Collusion and Predation Under (Almost) Free Entry', *International Journal of Industrial Economics*, 7: 381–401.
Hey, J. D. (1983), 'Towards a Double-Negative Economics', in J. Wiseman (ed.), *Beyond Postive Economics?*, London: Macmillan, 160–75.
Hirsch, F. (1977), *Social Limits to Growth*, London: Routledge and Kegan Paul.
Hirschleifer, J. (1985), 'The Expanding Domain of Economics', *American Economic Review*, 75: 53–68.
Hirschman, A. O. (1970), *Exit Voice and Loyalty: Responses to Decline in Firms, Organizations and States*, Cambridge, Mass.: Harvard University Press.
—— (1976), 'Discussion' (repr. as 'Exit and Voice: Some Further Distinctions'), *American Economic Review, Papers and Proceedings*, 66: 386–9.
—— (1981), 'The Social and Political Matrix of Inflation: Elaborations on the Latin American Experience', in *Essays in Trespassing: Economics to Politics and Beyond*, Cambridge: Cambridge University Press, 177–207.
—— (1982), *Shifting Involvements: Private Interest and Public Action*, Princeton, NJ: Princeton University Press.
—— (1985), 'Against Parsimony: Three Easy Ways to Complicating Some Categories of Economic Discourse', *Economics and Philosophy*, 1: 7–21.
Hirschmeier, J., and Yui, T. (1981), *The Development of Japanese Business 1600 –1980*, 2nd edn., London: Allen and Unwin.
Hodgson, G. (1986), 'Behind Methodological Individualism', *Cambridge Journal of Economics*, 10: 211–24.

Hodgson, G. (1988), *Economics and Institutions: A Manifesto for a Modern Institutional Economics*, Oxford: Polity Press.

Hogg, M. A., and Abrams, D. (1988), *Social Identifications: A Social Psychology of Intergroup Relations and Group Processes*, London: Routledge.

Jaques, E. (1982), *Free Enterprise, Fair Employment*, New York: Crane Russak; London: Heinemann.

Jones, S. R. G. (1984), *The Economics of Conformism*, Oxford: Blackwell.

Kerr, C., and Siegel, A. (1954), 'The Interindustry Propensity to Strike: An International Comparison', in A. Hornhauser, R. Dubin, and A. M. Ross (eds.), *Industrial Conflict*, New York: McGraw Hill, 189–212.

Kets de Vries, M. F. R. (1987), 'Prisoners of Leadership', INSEAD Working Papers 87–36, Fontainebleau.

Kirzner, I. M. (1973), *Competition and Entrepreneurship*, Chicago: University of Chicago Press.

Klamer, A. (ed.) (1984), *New Classical Macroeconomics: Conversations with New Classical Economists and their Opponents*, Brighton: Wheatsheaf.

Klein, B., and Leffler, K. (1981), 'The Role of Market Forces in Assuring Contractual Performance', *Journal of Political Economy*, 89: 615–41.

Knight, F. H. (1935), *The Ethics of Competition and Other Essays*, London: Allen and Unwin.

Kolm, S. C. (1969), 'The Optimum Production of Social Justice', in J. Margolis and H. Guitton (eds.), *Public Economics*, London: Macmillan, 145–200.

Kreps, D., and Wilson, R. (1982), 'Reputation and Incomplete Information', *Journal of Economic Theory*, 27: 253–79.

Lauterbach, A. (1954), *Man, Motives, and Money: Psychological Frontiers of Economics*, Ithaca, NY: Cornell University Press.

Leff, N. H. (1986), 'Trust, Envy, and the Political Economy of Industrial Development: Economic Groups in Developing Countries', Conference on The Role of Institutions in Economic Development, Cornell University, Ithaca, NY, mimeo.

Loasby, B. J. (1976), *Choice, Complexity and Ignorance*, Cambridge: Cambridge University Press.

Lodge, G. C., and Vogel, E. F. (eds.) (1987), *Ideology and National Competitiveness: An Analysis of Nine Countries*, Boston, Mass.: Harvard Business School Press.

Loomes, G. C., and Sugden, R. (1982), 'Regret Theory: An Alternative Theory of Rational Choice under Uncertainty', *Economic Journal*, 92: 805–24.

—— —— (1987), 'Some Implications of a More General Form of Regret Theory', *Journal of Economic Theory*, 41: 270–87.

Macaulay, S. (1963), 'Non-Contractual Relations in Business', *American Sociological Review*, 28: 55–70.

McIntosh, D. (1969), *The Foundations of Human Society*, Chicago: University of Chicago Press.

McKinley, E. (1965), 'Mankind in the History of Economic Thought', in B. F. Hoselitz (ed.), *Economics and the Idea of Mankind*, New York: Columbia University Press, 1–40.

McPherson, M. S. (1988), 'Reuniting Economics and Philosophy', in G. C. Winston and R. F. Teichgraeber III (eds.), *The Boundaries of Economics*, Cambridge: Cambridge University Press, 71–87.

March, J. G. (1978), 'Bounded Rationality, Ambiguity and the Engineering of Choice', *Bell Journal of Economics*, 9: 587–608.

Margolis, H. (1982), *Selfishness, Altruism and Rationality: A Theory of Social Choice*, Cambridge: Cambridge University Press.

Marshall, J. (1968), *Intention in Law and Society*, New York: Funk and Wagnalls.

Moore, B., Jr. (1978), *Injustice: The Social Bases of Obedience and Revolt*, London: Macmillan.

Morishima, M. (1982), *Why Has Japan 'Succeeded'?: Western Technology and the Japanese Ethos*, Cambridge: Cambridge University Press.

Mueller, D. C. (1990), *Public Choice II*, Cambridge: Cambridge University Press.

Myers, M. L. (1983), *The Soul of Economic Man: Ideas of Self-interest, Thomas Hobbes to Adam Smith*, Chicago: University of Chicago Press.

Naylor, R. (1987), 'Strikes, Free-riders and Social Customs', *Warwick Economic Research Papers*, No. 275.

North, D. (1979), 'A Framework for Analysing the State in Economic History', *Explorations in Economic History*, 16: 249–59.

Olson, M., Jr. (1965), *The Logic of Collective Action*, Cambridge, Mass.: Harvard University Press.

—— (1982), *The Rise and Decline of Nations*, New Haven: Yale University Press.

Ouchi, W. (1981), *Theory Z: How American Business Can Meet the Japanese Challenge*, Reading, Mass.: Addison-Wesley.

Parfit, D. (1978), 'Prudence, Morality, and the Prisoner's Dilemma', *Proceedings of the British Academy*, 65: 539–64, repr. in J. Elster (ed.), *Rational Choice*, Oxford: Blackwell, 1986, 34–59.

Pemberton, J. (1985), 'A Model of Wage and Employment Dynamics with Endogenous Preferences', *Oxford Economic Papers*, 37: 448–65.

Pen, J. (1985), *Among Economists*, Amsterdam: North Holland.

Pieters, R. G. M., and Van Raaij, W. F. (1988), 'The Role of Affect in Economic Behaviour', in W. F. van Raaij, G. M. van Veldhoven, and K.-E. Wärneryd (eds.), *Handbook of Economic Psychology*, Dordrecht: Kluwer Academic, 108–42.

Pollak, R. A. (1985), 'A Transaction Cost Approach to Families and Households', *Journal of Economic Literature*, 23: 581–608.

Porter, M. E. (1990), *The Competitive Advantage of Nations*, New York: Free Press.

Radner, R. (1985), 'Repeated Principal-Agent Games with Discounting', *Econometrica*, 53: 1173–98.

Rasmusen, E. (1989), *Games and Information: An Introduction to Game Theory*, Oxford: Blackwell.

Roberts, B., and Holdren, R. R. (1972), *Theory of Social Process: An Economic Analysis*, Ames, Ia.: Iowa State University Press.

Romer, D. (1984), 'The Theory of Social Custom: A Modification and Some Extensions', *Quarterly Journal of Economics*, 99: 717–28.

Russell, T., and Thaler, R. (1985), 'The Relevance of Quasi Rationality in Competitive Markets', *American Economic Reviews*, 75: 1071–82.

Schein, E. H. (1985), *Organisational Culture and Leadership*, San Francisco: Jossey-Bass.

Schelling, T. C. (1978*a*), 'Egonomics, or the Art of Self-Management', *American Economic Review, Papers and Proceedings*, 68: 290–4.

—— (1978*b*), *Micromotives and Macrobehaviour*, New York: W. W. Norton.

—— (1984), *Choice and Consequence: Perspectives of an Errant Economist*, Cambridge, Mass.: Harvard University Press. (Chapter 3: 'The Intimate Contest for Self-Command', 57–82).

Scitovsky, T. (1976), *The Joyless Economy*, New York: Oxford University Press.

Sen, A. K. (1977), 'Rational Fools: A Critique of the Behavioural Foundations of Economic Theory', *Philosophy and Public Affairs*, 6: 317–44, repr. in *Choice, Welfare and Measurement*, Oxford: Blackwell, 1982, 84–106.

—— (1983), *Poverty and Famines: An Essay on Entitlement and Deprivation*, Oxford: Oxford University Press.

Sen, A. (1987), *On Ethics and Economics*, Oxford: Blackwell.

Simon, E. (1972), *The Anglo-Saxon Manner: The English Contribution to Civilisation*, London: Cassel.

Simon, H. A. (1983), *Reason in Human Affairs*, Oxford: Blackwell.

Smith, A. (1759), *The Theory of Moral Sentiments*, ed. D. D. Raphel and A. L. Macfie, Oxford: Clarendon Press, 1976.

—— (1776), *An Inquiry into the Nature and Causes of the Wealth of Nations*, ed. R. H. Campbell, A. S. Skinner, and W. B. Todd, Oxford: Clarendon Press, 1976.

—— (1795), 'The Principles which Lead and Direct Philosophical Enquiries; Illustrated by the History of Astronomy', in *Essays on Philosophical Subjects*, ed. W. P. D. Wightman, J. C. Bryce, and I. S. Ross, Oxford: Clarendon Press, 1976, 31–105.

Smith, J. M. (1982), *Evolution and the Theory of Games*, Cambridge: Cambridge University Press.

Sugden, R. (1986), *The Economics of Rights, Co-operation and Welfare*, Oxford: Blackwell.

Tainter, J. A. (1988), *The Collapse of Complex Societies*, Cambridge: Cambridge University Press.

Thaler, R., and Shefrin, H. M. (1981), 'An Economic Theory of Self-control', *Journal of Political Economy*, 89: 396–406.

Thurow, L. C. (1983), *Dangerous Currents: The State of Economics*, Oxford: Oxford University Press.

Trivers, R. L. (1971), 'The Evolution of Reciprocal Altruism', *Quarterly Review of Biology*, 46: 35–57.

Vance, N. (1985), *The Sinews of the Spirit: The Ideal of Christian Manliness in Victorian Literature and Religious Thought*, Cambridge: Cambridge University Press.

Vanek, J. (1971), *The Participatory Economy: An Evolutionary Hypothesis and a Strategy for Development*, Ithaca NY: Cornell University Press.

Weber, M. (1947), *The Theory of Social and Economic Organisation* trans. A. M. Henderson and T. Parsons, ed. T. Parsons, New York: Oxford University Press.

Whitley, R. (1984), *The Intellectual and Social Organization of the Sciences*, Oxford: Clarendon Press.

Williamson, O. E. (1985), *The Economic Institutions of Capitalism: Firms, Markets, Relational Contracting*, New York: Free Press.

Wolinsky, A. (1987), 'Information Revelation in a Market with Pairwise Meetings', *Warwick Economic Research Papers*, No. 284.

Index

abilities: differential in manipulation strategy 44–6
Abrams, D. 233
achievement
  inspiration of 29–52
  motivating 30–3; and culture of firms 233–5
Akerlof, G. A. 27, 67, 100
Alchian, A. A. 135, 233
allegiance
  competition for 229–31
  and loyalty 232
altruism: defined 266
anger 124–7
  in encounter decisions 258
  mechanism of 125
announcement
  and crime rate 86
  effect of 76–80
  in manipulating team effort 139–40
  optimal strategies for 77–8; value of 79
  as substitute for manipulation 77
arousal
  encounter as three-stage game 126
  see also anger
Arrow, K. J. 177
Assurance (game) 55–7, 67
  optimal strategies for announcement 78
  teamwork as 133–4
asymmetric co-ordination gains in pairwise encounters 59
Austrian school of supply side economics 245
autonomy of individual, rejected 23–4
avoidance
  and cheating in encounter decisions 257
  encounter as three-stage game 126
  and follower's data set for participation 108–10
  incentives for and personal reputation in small groups 175–6
  and reciprocity 118–19
  satisfaction from reciprocity 122
  of trade 106
Axelrod, R. 23, 27

Barrientos, A. 23
Barry, B. 23
Bartlett, R. 23
Baxter, J. L. 23
Becker, G. S. 24
Becker, L. C. 116
behaviour, norms of 20–1
Blois, K. J. 239
Bloom, A. 263
Booth, A. L. 214
broad materialism
  defined 266
  and free rider problems with private goods 197
  and leader's preferences 33–4
  in trading encounters 65
  welfare implications of team production 139–40
Brubaker, R. 13
Buchanan, J. M. 90
business culture 225–42
  groups, typology of 226–9
  internal organization of firm 260–2
  and leadership 261–3
  leadership styles 235–7
business education 262–3

Campbell, T. D. 23
cartels 212–14
  information set for 215; distributional effects of cheating 215–17
  optimal manipulation in 214–18
Casson, M. C. 23, 233, 239
cheating 55
  defined 266
  emotional rewards of 60
  encounter as three-stage game 126
  and follower's data set for participation 108–10
  follower's perceived rewards for participating in trade under guilt 171–2
  and guilt in manipulation strategies 256
  incentive to 59–62
  and information set for cartels 215–17; distributional effects 215–16

cheating (*cont.*)
information sets for middlemen and customer 152–3; middlemen cheating 155–6; monitoring by 157–62
material rewards of 59
by middlemen 154–62
mutual, equilibrium of 91; stability of 93
and participation 100–1, 257
and personal reputation in small groups 172–4; incentives for honesty 175–6
for private good provision 195–6
and public good provision 192–4
and reciprocity 116, 118–20
and revenge 127
satisfaction from reciprocity 124
and trader's data set for participation 102–4
in trading encounters 63–5
Chell, E. 169
Chicken (game) 55–7, 67
optimal strategies for announcement 78
choice: continuity of, rejected 25
Church, R. 234
club as group 227, 228
co-operation
inter-firm 7–9, 238–9
inter-group 237–9
co-ordination
defined 266
and leader's preferences 33
multilateral 137
Coase, R. H. 26
Cochran, T. C. 262
Cohen, M. D. 23
collaboration and inter-firm co-operation 8–9
Collard, D. 197
collective negotiations, solidarity in 213–14
collusion
and culture of firms 233–5
as group activity 213
and groups 219–21
groups for 226
and inter-group co-operation 239
and loyalty 222
and monitoring 214
as moral crusade 213–22
and size of groups 218
communication costs
of alternate strategies 177
by dispersion of group 178
least-costs strategies 177
in small groups 176–80
comparative cultural analysis 3–11
inter-firm co-operation 7–9
personnel policy 4–7
urban growth and decline 9–11
competition
for allegiance 229–31
and distrust in United Kingdom 252
economic freedom and 246–8
influence on leadership 248; and groups 248–9
and justice 202–3
and manipulation in cartels 216
and middlemen 165–6
and monopoly 247
between rival cultures 239–41
and selfish leaders 201–202
contractarian ethics
and distributive justice 234
to distributive justice 208–9
and firms' activities 231
Cournot follower and Stackelberg leader 36
crime rate
defined 266
dynamics in repeated encounters 86–9
dynamics of with repeated operation 94
in encounters 61
equilibrium of 91–2
and holistic approach to distributive justice 209
impact of intensity of manipulation on 69–71, 73–4, 76
impact of size and dispersion of group 181
and independent slacking decisions 141
leader's announcement of 76–7
and manipulation strategy 36, 39, 45
and personal reputation in small groups 173
and reciprocity 120–1
in trade encounters 63–5
and voluntary participation 100
cultural engineering for economic policy 243–54
culture
competition between 239–41
diversity 240
of economics profession 22–3
export of 240

and trade 55–81
customer
cheating: information set and reward for 155–6; monitoring by 157–8
information set and reward for 152

Dahrendorf, R. 18
de-skilling and personnel policy 6–7
decentralization and supply-side economic policy 244–6
dedication
defined 266
in monitoring achievement 47
and motivating achievement 31–3
and rewards 32
and teamwork 137–9
see also honesty
Demsetz, H. 135, 231
dispersion
costs of alternative communication strategies 177; least-cost 177; manipulation, by 177
impact of on co-ordination strategy 180
and small groups 176–80
distribution
and justice 200–212
and welfare 206–7
distributive justice 203–7
and contractarian approach 207–8
and contractarian ethics of firm 234
and disincentive effects of taxation 208
and moral manipulation 210–12
distrust
effects in United Kingdom 249–52
influence on leadership 248–9; and monitoring 248
and transaction costs 255
Drakopoulos, S. 23

Earl, P. E. 23, 240
economic activity as encounters between individuals 13–14
economic collapse and holistic approach to distributive justice 209–10
economic freedom in competitive economy 246–7
economic groups 227
economic man
and ethical man 23–6
and nature of trust 15–17
economic performance
microfoundations of 12–14
and participation 100
and transaction costs 3
economic policy
implications for 243–54
supply-side 243
supply-side economics, *see* supply-side economic policy
in United Kingdom 249–52
economics
culture of profession 22–3
and groups 232–3
groups, function of 226–7
and learning 84
effort
cost of in manipulation strategy 41
technology 135; and teamwork 137
see also marginal productivity of effort
Elster, J. 27
emotional rewards
of cheating 59–60
of honesty 59–60
in motivating achievement 32
and participation 122
and reciprocity 118–19, 122
encounters
emotional rewards from 60
and groups 226–8
between individuals, economic activity as 13–14
material rewards from 59–60
modelling of 57–9
monitoring of 16–17
moral mechanism of 16–17
multi-stage decisions 257–9
one-off: contrasted with repeated encounters 82, 85; general theory of 67–76; and participation 108–11
teamwork as special type of 135
as three-stage game 126
trade as 58–9
see also pairwise encounters; repeated encounters; trading encounters
engineering technology 135
enlightened self-interest and nature of trust 16–17
environment
concerns in United Kingdom 253–4
perception of 24
ethical man and economic man 23-6
Etzioni, A. 27

family
and cultural environment of firm 236–7
as group 227

fear as leadership style 235–7
Felix, D. 23
firms
  culture of 232–5; and collusion 234; and loyalty 234–5; and private property 233; and teamwork 233
  as groups 227, 228
  and inter-group co-operation 239
  internal organization of 260–1
follower
  data set: for effort choice 31–2; for encounters 60; undermonitoring 47; for private good provision 195–6; for public good provision 192–4; reciprocity 120–1; satisfaction from reciprocity 122; subjected to lump-sum tax 205–6; for trading encounters 63; and voluntary participation 108–9
  defined 264
  perceived rewards for participating in trade under guilt 171–2
  strategy of
    and emotional rewards 60
    and material rewards 59
free-rider
  and moral manipulation 192
  and philanthropy 189–200
  and private goods 195–7
  and public goods in economic theory 190–1
Freeman, R. B. 214
Friedman, J. W. 26, 82
Fudenberg, D. 169
Furnham, A. 27

Gambetta, D. 15
games 55–7
  theory, conventional and rewards 57–8
  three-stage, encounter as 126
  and voluntary participation 100
  see also Assurance; Chicken; Harmony; Prisoners' Dilemma
groups
  affiliation 230–1
  allegiance in 229–31
  co-operation between 237–9
  and collusion: and loyalty 221; and redistribution 219–21; size of 218
  function of 226–7
  impact of factors in manipulation strategy 41
  income, distribution of in 189–90
  and intermediators 233
  leader and competition 248–9
  and loyalty 231–32
  quitting, risk of 231
  reputations of 230–1
  and sensitivity 232
  and team production 233
  typology of 226–8
  see also large groups; small groups
guilt
  alternative patterns of 116–17
  defined 266
  in encounters 61
  follower's perceived rewards for participating in trade under guilt 171–2
  intensity of manipulation on crime rate and 68–70, 71
  in manipulation strategies 256
  and motivating achievement 31
  and reciprocity 121
  and satisfaction 125
  in trade encounters 63
  and voluntary participation 108

Hamlin, A. P. 202
Haplerin, R. H. 13
Hargreaves-Heap, S. 24
Harmony (game) 55–7
Harrington, J. E. 213
Hey, J. D. 24
Hirsch, F. 13
Hirschleifer, J. 22
Hirschman, A. O. 27, 231
Hirschmeier, J. 234
Hodgson, G. 25, 26
Hogg, M. A. 233
Holdren, R. R. 226
holistic approach to distributive justice 208–30
honesty 55
  and announcement of crime rate 77
  defined 266
  emotional rewards of 60
  encounter as three-stage game 126
  and follower's data set for participation 108–10
  follower's perceived rewards for participating in trade under guilt 171–2
  incentives for created by reputation in small groups 175–6

and information set for cartels 213
information sets for middlemen and customer 152–3; middlemen cheating 155–6; monitoring by 157–62
material rewards of 59
mutual, equilibrium of 91; stability of 93
for private good provision 195–6
and public good provision 192–4
and reciprocity 118–19
and revenge 127
satisfaction from reciprocity 122
and trader's data set for participation 102–4
in trading encounters 63–5

hostages
and collusion 218–19
and intermediators 149, 158
and manipulation strategies 46–7
and nature of trust 16

incentives
created by reputation in small groups 175–6
and disincentive effects of taxation 207–8
nature of problem 34–5
team-based 142–3

income, distribution of in group activity 189–90

individuals
autonomy of, rejected 23–4
and competition for allegiance 230
economic activity as encounters between 13–14
and groups 228–9
and guilt, sensitivity to 31
optimism of 20–1
trustworthiness of 19–20
utility function, fixed 24

information
and leadership 258–9
and learning 84–5
and reputation 170–1
and reward set for cheating middlemen 155–6

injustice
and anger 125

innovative industry
and personnel policy 7
and training 11–12

inter-firm co-operation
in comparative cultural analysis 7–9
and trust 7–9

intermediation: process of 152

intermediators 148–68
as class 151
and groups 231
and hostages 149, 158
and leaders 148–9
as reputable but selfish 150–4

intuition: impact of factors in manipulation strategy 41

Japan: economic culture 243–4

Jacques, E. 234

Jones, S. R. G. 27

justice
distributive 203–7
psychological need for 202–3

Kerr, C. 234

Kets de Vries, M. F. I. 236

Kirzner, I. M. 245

Klamer, A. 22

Klein, B. 16

Knight, F. H. 235

Kolm, S. C. 203

Kreps, D. 27, 169

Laffer curve for taxation 208

large groups
and collusion 220–21
and communication costs 178–80

Lauterbach, A. 18

leader
achievement: inspiration of 29–52; motivating 30–3
altruistic: objectives 33–4; and participation 105; and personal reputation in small groups 174, 176; and repeated encounters 89; and welfare 206–7
charisma, degree of 41
commitment by 18–19
competition: and groups 248–9
defined 266
economic collapse and holistic approach to distributive justice 209–30
impact of factors in manipulation strategy 41
incentive problem, nature of 34–5
and intermediators 148–9
and middleman who cheats 156

leader (*cont.*)
monitoring: as alternative to manipulation 46–8; and distrust 248–9
optimal manipulation strategy 35–8; comparative statics 38–42; differential abilities 44–6; monitoring as alternative 46–8; relative transaction costs 48–51; sensitivity analysis 42–4
policy objectives: control of 33–4; and narrow materialism 45
power, asymmetry of 210–11
and repeated encounters 89
and revenge 117
self-disciplined 209
selfish: and competition 201–202; and distributive justice 203–4; equity and material performance 206–7
and team effort 136–7
utility and narrow materialism 45
weak and costs of manipulation 235–6
leadership
announcement of crime rate, effect of 76–7
and business culture 261–3
and distrust 248–9
and division of labour 17–18
and information 258–9
and monitoring of encounters 18
moral, advantages of 255–6
moral rhetoric of 20–1
styles 235–7
team, and middlemen 162–5
techniques of 19–22
and voluntary participation 101
learning
from experience 84–5
and repeated encounters 82–4
Leff, N. H. 3–4
Leffler, K. 16
Lewis, A. 27
Loasby, B. J. 24
local community as group 227
Lodge, G. C. 23
Loomes, G. C. 20
loyalty
and collusion 221
and culture of firms 233–4
defined 266
and groups 231–32
and sensitivity 232

Macaulay, S. 239
McIntosh, D. 18
McKinley, E. 18
McPherson, M. S. 18
manipulation
announcement a substitute for 77
and business culture 261–63
of contribution for public goods 91–94
costs of 35; and leader's impact 41; and repeated trading encounters 106–7; in trading encounters 63, 66; and voluntary participation 100–1; weak leaders 235–6
diminishing returns to, optima for 72
gains from 75
intensity of: and cartels 216–17; of contribution for public goods 194; and group size 172; impact on crime rate 69–71, 73–4, 76, 86–7; and learning 85; and narrow materialism 45; and philanthropy 198; and team effort 141
and inter-group co-operation 238–9
and intermediation 153–4
marginal cost of: and cartels 217; and middleman's cost of manipulation 159; and size of group 181
and monitoring: as alternative to 46–8; impact of size and dispersion of group 180–3; in trading encounters 66–7
optimal strategy 35–8; and cartels 214–18; comparative statics 38–42; comparison of strategies 43; conclusions 259–60; differential abilities 44–6; monitoring as alternative 46–8; relative transaction costs 48–51; and repeated encounters 98–90; sensitivity analysis 42–4
and optimal team size 143–6
and Prisoners' Dilemma 63
and satisfaction 124
strategies, summary of 256–7
and team-based incentives 142–3
of team effort 136–40
see also moral manipulation
manufacturing industries: large-scale and personnel policy 6–7
marginal productivity of effort
changes in and manipulation strategy 39–41
defined 264

and hostages 47–8
and optimal team size 145–6
Margolis, H. 195
Marshall, J. 21
material rewards
of cheating 59
of honesty 59
in monitoring achievement 47
in motivating achievement 32
materialism: defined 266
Medoff, J. L. 214
middle class: intermediators as 151
middlemen
cheating 154–62; information and reward set for 155–6, monitoring by 157–8
and competition 165–6
information and reward set for 152
intermediators as 151
moral manipulation by 149, 223
and team leadership 162–5
monitoring
cheating middlemen, information and reward set for 157–8
and collusion 214; as alternative in 216–17
conclusions 259–60
and effects of distrust in United Kingdom 251
of encounters 16–17
follower's data set under 47
inputs and outputs in trade 58–9
and intermediation 154
and leadership styles 235–6
and manipulation: as alternative to 46–8; impact of size and dispersion of group 180–3; in trading encounters 66–7
and private goods 194–7
of public goods 191
and repeated encounters 90–1
and team-based incentives 142–3
monopolist, intermediators as 150–4
Moore, B. 202
moral manipulation
and announcements 77
and collusion 219
and distributive justice 208–9
and firms 233, 237
and free-rider problem 194
and middlemen 149, 225
and philanthropy 197
in small groups 171–72, 181
and team size 145
moral mechanism of encounters 16–17
moral rhetoric of leadership 20–1
morality: social dimension of 255
Morishima, M. 23
Mueller, D. C. 187
multi-stage decisions in encounters 257–9
multilateral co-ordination 137
Myers, M. L. 23

narrow materialism
defined 266
and free rider problems with private goods 196–7
and leader's preferences 33–4, 44–5
in trade encounters 61–2, 65
nation-state
and cultural environment of firm 234
as group 227, 228–9
nationalism and unity 252–4
Naylor, R. 27, 214
non-co-operative game: defined 266
notation 264–5

Olson, M. 23, 190
Ouchi, W. 235

pairwise encounters 58
two-person teamwork as 55, 133
Parfit, D. 17
participation
encounter as three-stage game 126
one-off encounters 108–11
promoting 100–15
repeated encounters: general case 111–15; in trade 106–8
in trade: avoidance of 106; incentives 102–6; repeated encounters 106–8
voluntary 100–2
Pemberton, J. 27
Pen, J. 22
personal reputation in small groups 170–4
incentives created by 175–6
personnel policy
in comparative cultural analysis 4–7
and trust 4–6
philanthropy 195–6
and free-rider problem 189–200
groups for 226
Pieters, R: G. M. 27
policy objectives: leader's control of 33–4
Pollak, R. A. 226

Porter, M. E. 239, 247
power: asymmetry of leader's 210–11
preferences of leader and co-ordination 33
Prisoners' Dilemma (game) 55–7
  defined 266
  and manipulation in cartels 216
  as manipulation strategy 256
  and philanthropy 198
  and role of manipulation 63
private goods
  and monitoring 194–7
  see also public goods
product life-cycles
  and personnel policy 7
  and trust 11–12
production: groups for 226
public choice as economic policy 189
public goods
  in economic theory 190–1
  groups for provision of 226
  manipulation of contributors 190–94
  see also private goods

Raaij, W. F. 27
Radner, R. 169
Rasmusen, E. 26
reciprocity
  and cheating 116, 118–20
  diagrammatic analysis 117–21
  and satisfaction 120–4
  and voluntary participation 120–1
recurrent encounters, *see* repeated encounters
redistribution between factions 219–21
religion
  and cultural environment of firm 236
  as group 227, 237
repeated encounters 82–4
  and crime rate 86–9
  equilibrium analysis 91–2
  instability of outcome 92
  and learning from experience 84–5
  and monitoring 90–1
  optimal manipulation with 89–90, 95
  and participation 101; general case 111–15; in trade 106–8
  and reciprocity 120
  stability analysis 92–8
  in trade: and participation 106–8; voluntary participation 114–15
resignation, and encounter as three-stage game 126
revenge
  and anger 124–7
  desire for and leader 117
  encounter as three-stage game 126
  rewards associated with 126
rewards
  associated with revenge 126
  in conventional game theory 57–8
  and dedication and slacking 32
  follower's perceived for participating in trade under guilt 171–2
  information sets for middlemen and customer 152–3; middlemen cheating 155–6; monitoring by 157–62
  non-pecuniary and groups 231
  and revenge 127
  see also emotional rewards; material rewards
Roberts, B. 226
Romer, D. 27
Russell, T. 27

satisfaction
  and reciprocity 122–4
  and vengeance 127–8
Schein, E. H. 169, 233
Schelling, T. C. 27
Scitovsky, T. 27
selectivity: and voluntary participation 101
self-discipline, morality of 210–11
selfishness
  intermediators as reputable but 150–4
  leader: and competition 200–2; and distributive justice 203–4; equity and material performance 206–7
  and supply-side economic policy 246
Sen, A. K. 17, 24, 199
sensitivity
  analysis in manipulation strategy 42–4
  in comparative statics 40
  defined 267
  and groups 232
  impact of intensity of manipulation on crime rate and 68–70, 71
  in intermediators 150–1
  and loyalty 232
  and manipulation in cartels 216–17
  in trade encounters 61
service sector and personnel policy 7
Shefrin, H. M. 27
Simon, E. 210

Simon, H. A. 57
size
  impact of on co-ordination strategy 180
  and small groups 176–80
slacking
  defined 267
  independent decision to 140–1
  in monitoring achievement 47
  and motivating achievement 31–3
  and rewards 32
  and teamwork 137–9
  see also cheating
small groups 169–84
  and collusion 219–21
  communication costs 176–80; of alternate strategies 177
  personal reputation in 170–4; incentives created by 175–6
  size and dispersion 176–80; impact on choice of co-ordination, strategy 180–3
  transaction costs in 169–70
Smith, A. 23, 136
Smith, J. M. 26
social groups 227
society, co-ordination within 25
solidarity
  in collective negotiations 213–14
  and information set for cartels 215
Stackelberg leader 25–6
  defined 267
  leader as 36
Stigler, G. J. 24
Sugden, R. 14, 20
supply-side economic policy 243
  and competition 246
  and decentralization 244–6
  and human nature 244
symmetric co-ordination gains in pairwise encounters 59

Tainter, J. A. 209
taxation
  disincentive effects of 207–8
  and distributive justice 204–6
  Laffer curve for 208
  and public goods 191
teams
  and collusion 213
  effort: manipulation of 136–40; and multilateral co-ordination 137
  groups as 226
  incentives 142–3
  and leadership: information set for 163; and middlemen 162–5
  members: data set for production 138; information set for 163
  production: and groups 233; member's data set for 138; welfare implications of 139–40
  size: and effort technology 135–6; and engineering technology 135–6; optimal 143–6
  spirit 133–47
teamwork
  complementary efforts in 135
  as pairwise encounter 133–4
  technologies of 135–6
Thaler, R. 27
trade
  economy of 62–6
  as encounter 58–9
  follower's perceived rewards for participating in under guilt 171–2
  groups for 226
  and intermediators 148–9
  participation: avoidance of 106; incentives for 102–6; and repeated encounters 106–8; voluntary, equilibrium and stability of 114–15
  see also repeated encounters
trade unions
  effects of distrust in United Kingdom 249–52
  as groups 227, 228
trader: data set for participation 102
trading encounters, strategies in 62–6
training
  and inter-firm co-operation 8–9
  and personnel policy 6–7
  and product life-cycles 12
transaction costs
  and distrust 255
  and economic performance 3, 225
  emphasis on in other theories 26
  individual in group, replacing 231
  and inter-group co-operation 238
  and intermediators 148
  and manipulation 90–1
    in trading encounters 66–7
  and monitoring 91
  of monitoring 48–51
  and optimal team size 145–6
  relative in manipulation strategy 48–51
  in small groups 169–70
Trivers, R. L. 169

trust
  engineering of 20–1
  and holistic approach to distributive justice 209
  and inter-firm co-operation 7–9
  and leadership 248–9
  as leadership style 235–7
  nature of 15–17
  and personnel policy 4–6
  political economy of 11–12
two-person teamwork: pairwise encounters as 55, 133

United Kingdom
  economic culture 243–5
  effects of distrust 249–52
  industrial relations 245
United States
  business leadership 261–2
  economic culture 243–4
  supply-side economics 244
unity: national missions of 252–4
universal consultation 202–3
urban growth and decline in comparative cultural analysis 9–11
utilitarianism
  defined 267
  and leader's preferences 34
utility as material component in economics 24
Vance, N. 210
Vanek, J. 202
vengeance
  encounter as three-stage game 126
  in encounter decisions 258
  moral limits to 127–8
Vogel, E. F. 23
voluntary participation 100–2
  equilibrium and stability of recurrent trade 114–15
  and reciprocity 120–1

Walrasian auctioneer in supply-side economics 245
Weber, M. 13
welfare concerns in United Kingdom 253–4
Whitley, R. 22
Williamson, O. E. 26
Wilson, R. 27
Wolinsky, A. 13
work ethic
  and leader's preferences 34
  in trading encounters 65

Yellen, J. L. 27
Yui, T. 234